The Secret Code of PM

The Secret Code of PM

DJ Dromgold

First published 2016
Alexander Publications

Copyright © 2016 DJ Dromgold

This is a work of fiction. Names, characters, businesses, places, events and incidents are either the products of the author's imagination or used in a fictitious manner. Any resemblance to actual persons, living or dead, or actual events is purely coincidental.

Typeset by BookPOD Pty Ltd

ISBN: 978-0-9953567-0-2 (pbk)
eISBN: 978-0-9953567-1-9

A Cataloguing-in-Publication record is available
from the National Library of Australia

Med•evolv

A start-up business
(aka a high risk project)

Organization chart

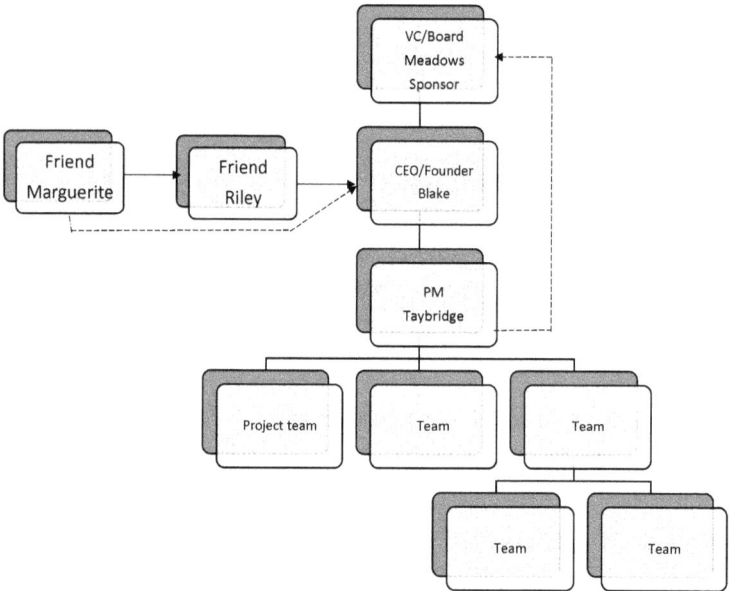

$\mathcal{O}$NE

'What is he hiding from me?'

Blake stared at the data on his computer screen. Aiden Taybridge, his project manager, sent him status reports at midday every Friday, on the dot. Taybridge's reports were big on metaphor, low on detail. On the left-hand side of each page was a series of dials, like the water temperature gauge on Blake's BMW. The colors on each gauge shifted from green to orange to red. Green meant the engine—sorry, one element of the project—was running smoothly, on target. Orange meant a minor issue, but keep your foot down, Taybridge has it all under control. Red meant a major problem that demanded Blake's attention. Pull off the road before the engine seizes. The status report was easy to visualize, easy to understand. And, from Blake's point of view as the CEO of Med•evolv, deeply patronizing.

But the colorful reports did not stop there. On the right-hand side of the page, each project element had been allocated its own colored box, along with a line or two of explanatory text. The boxes were stacked on top of each other, like traffic lights. As Blake flicked through the document, all he saw were green icons, with a smattering of orange. The orange boxes contained scant information: *Minimal overrun, low risk.* The entire report seemed to have only one purpose: to reveal nothing.

'Is this a project report or a promo for a Pixar movie?' Blake asked himself. 'I need information I can use—not a parody of *Cars*!'

Blake felt the muscles tightening in his neck, warning of an impending migraine. He reached for the ibuprofen. As he washed the tablets down with a glass of water, his anxiety increased.

He knew all too well the risks involved with this project. You don't raise $200 million in venture capital to deliver something easy. Med•evolv dreamed big. They were playing for high stakes. Nothing less than a complete disruption to the field of human genome sequencing. This project pushed the boundaries of science, IT, and the health insurance industry. Blake knew they would encounter problems—massive problems. But Taybridge's report made it all seem as easy as cruising down the Pacific Coast Highway.

'I'll find out what's really going on,' he pledged, 'even if it does my head in! But I'll need some help.' He picked up his phone, and called Riley.

Blake tapped his fingers on the table and looked around the coffee shop, as it began to fill with the mid-morning rush. 'I'm looking for some kind of narrative that gives me an honest appraisal of our project status,' he began. 'Can't find it! All I see are these little slivers of work that are signed off. Done. Complete. Perfect. But when I interrogate the report, none of these little slivers address the crucial problems we need to solve.'

Riley frowned. 'Before we begin, why don't you order me a double-shot ristretto latte? Once I know that's been taken care of, I'll be able to give you my full attention.'

Blake offered her a rueful smile, and went up to the counter. 'Never put yourself between Riley and her coffee,' he reminded himself. He ordered a mocha for himself—even though the caffeine and sugar seemed to feed his headache.

'Can you give me an example of these crucial problems you need to solve?' Riley asked when he returned to their corner table.

Blake smiled. Not an easy smile. More of a grimace. 'You know I can't tell you. Nondisclosure and all that.'

'How long have we known each other?'

'Since kindergarten,' Blake admitted.

'We've gone through school and college together. Whatever direction life has taken us, we've always complemented each other. You used to say that if—'

'If I had a twin sister, she'd be you.' Blake's smile relaxed this time. 'That's true. But if I tell you about this project, I guess I'm worried you won't approve.'

'Of you? Or of the project?'

'Both. I feel so strongly identified with this one. I can't separate myself out from it.'

'So it matters deeply to you. And Taybridge, from your point of view, isn't delivering what you need.'

'True true.'

'Does he know what you expect?'

'Of course. We've been through a heap of scoping documents and endless meetings. It's all there.'

'How long are those scoping papers?'

'All told? Thousands of pages.'

'It's a complex project.'

'Very complex.'

'No doubt. But does Taybridge understand the essence of it?'

'He should do. Where are you going with these questions?'

'Trying to get to the nub of your problem. Remember that robotics project we worked on during our first year at college?'

'Don't remind me. All these years I've been trying to forget.'

'Because we only came second?'

'Like I said, I'm trying to forget.'

'OK, Blake. Don't be any harder on yourself than you need to be. One evening we were up in your dorm room with the other three guys on the team, planning the project. We all knew you were leading us. You told us what you wanted to achieve. Ring any bells?'

'Yes. I said I wanted to blow the Dean's socks off. I wanted him to come up to us and say, "You guys rock!"'

'And what happened?'

'We came second to that Pushpaw team. What a bunch of nerds!'

Riley sighed. 'And the Dean, Blake?'

'He came up to us and said, "I don't know what you've done, but I can't take my eyes off that robot of yours. When you measure the contest against all the criteria, the

Pushpaw team scored a fraction higher, but your robot—it almost seems to have a mind of its own!" And then he went like this...' Blake pressed his fist against Riley's fist.

'Right,' Riley said. 'After which the Dean invited us to meet with him. After all, we were the ones who designed the software. Somehow, we'd stumbled across a set of algorithms that allowed us to build a digital gyroscope that gave the robot an almost intuitive sense of balance and direction. The three of us worked the last few bugs out of the system. And then, with the Dean's help, we patented the design...'

'And licensed it to Pyrouette for a mid-seven figure fee.'

'And the Pushpaw team?'

'Still subsisting on ramen noodles and tooth-rot cola, as much as I care. Mind you, that's not the full story...'

'Let's not go there right now,' Riley suggested. 'So if you could place Taybridge on our team or with the Pushpaws, where would he be?'

Blake sighed. 'It's not so simple now. Even a few million won't seed the kind of start-up I envisage. So I need venture capital, and my investors demand a professional project manager. They see me as just a kid, as too big a risk. So they foisted Taybridge on me.'

'He's part of their investment?' Riley asked.

'Their investment is conditional on his involvement.'

'Well then,' Riley said, 'you have a problem. A problem we are going to solve!'

The waitress brought their coffees. Morosely, Blake stirred another sachet of sugar into his mocha. That whole deal with Pyrouette tasted bitter in his mouth.

Riley jolted him back to the present. 'I don't understand why you insist on drinking coffee when you have a migraine coming on,' she said.

'We've been through this so many times, Ms. Pearce,' Blake replied. 'No matter how much pain these migraines cause me, I can't live without them. They're how I solve my problems. As they're coming on, I sense the answer is on the tip of my tongue. And as the migraine lifts, there it is, in plain sight—the perfect solution to whatever's been worrying me.'

'I know all that,' Riley said. 'It's just painful to watch.' Deftly, she changed the subject. 'So tell me. How hands-on have you been with the project?'

'Not so much. I've been relegated to more publicity roles.'

'That's got to change.' The expression on Riley's face made it clear she was not for turning. 'The reason we successfully patented that intuitive gyroscope was in no small way due to your leadership, your direct involvement. Whatever you're doing at Med•evolv, it won't succeed without your stewardship.'

'Why's that?'

'You want me to play to your vanity?'

'Not at all. It's a real question. I don't have the expertise needed to run this project. It's terribly complex. And I'm not that good with people. Mild Asperger's, remember?'

Riley nodded. 'We're going to fix that for you, Blake. When I joined Bank Pacific West, I met a consultant named Marguerite. The bank engaged her to set up our business-to-business payments project. There was a simple reason for this: in the years before I joined, the bank had run three consecutive projects, each of which failed spectacularly. They achieved nothing, other than burning tens of millions of dollars on executive follies.

'Marguerite had acquired a strong reputation as a go-to woman for complex projects, especially when all else had failed. As it turned out, her intervention at the start of the project made the difference between success and failure. Mind you, she put a few people's noses out of joint. She's quite unorthodox—which is part of her charm. In some ways she reminds me of you.'

'Except for the charm, which I lack, apparently.'

'You can be charming enough, Blake, when the mood takes you. How many of your employees followed you to Med•evolv?'

'Roughly half,' Blake said.

'So charm is not your problem,' Riley stated. 'Your problem's more profound than that. You're too intelligent.'

'In my book, there's no such thing as *too intelligent*,' Blake countered.

'But there is, Blake. Some people are so intelligent they believe they have nothing more to learn.' Riley lowered her voice. 'Remind you of anyone you know?'

Blake grimaced. 'So you're saying this Marguerite could help me, if only I'd let her?'

'She has the most spectacular array of piercings, Blake. She's our age—she gets what we're trying to achieve. It's not like she's going to talk down to you. She works with people, not against them.'

'And all the noses she put out of joint?'

'They were all people who challenged her. Not that she smacked them down—she simply showed them the error of their ways. Men of a certain age can't stand being outmaneuvered by a woman fifteen years their junior.'

'How come she knows so much if she's so young?'

'How do you rate yourself as an engineer and coder?' Riley asked.

'I'm in the top 2%,' Blake replied, without so much as blushing.

'How come you know so much if you're so young?' Riley smirked. 'Marguerite holds Master degrees in psychology and project management. She earned a Trans-Pacific PhD through the University of Sydney and Berkeley. Her topic? Project defeats and victories in government, IT, and financial services. Half the projects she studied failed big time; the other half transformed entire industries, showering their instigators with fame and money. None of these projects were more than two years old. Marguerite wanted currency, not ancient history. Before long she identified the main prerequisites for success, and worked out how to salvage even the most hopeless projects. Ironically, her timeline for her PhD slipped because she was winning so much consulting work. Her supervisor didn't mind because the projects she worked on provided further metrics to validate her thesis.

'But it's not Marguerite's doctorate that makes the difference. Other consultants hold top-flight academic qualifications. Where she differs is in her approach to everything: ***always question orthodoxy***. You can offer her all the Kool-Aid you want. She won't drink it.'

'Give me an example,' Blake demanded.

'OK. I have her Keynote presentation on my tablet.' Riley tapped the screen quickly. Here's her first point—the one she used to capture our full attention. *85% of projects fail.'*

'85%?' Blake exclaimed. 'That can't be right!'

'Sadly, it is. Marguerite has researched the history of project management. It's a history of increasingly stringent protocols, which led to decreasing success rates. Back in the sixties, maybe 50% of projects failed. In an effort to stem what was thought to be an unacceptable failure rate, the project management profession adopted more and more prescriptive procedures. But the results were surprising. Success rates plummeted. 85% of projects now fail to meet the expectations of their sponsors—the people who pony up the cash to run the project in the first place. Why is that?'

Blake shrugged his shoulders. 'Because people like Taybridge became project managers?'

'Perhaps. Or maybe they fail because people like you refused to step up to the mark.'

Blake could not help himself. 'You want me to manage this project? But project management is boring!' he blurted out.

'Tell me, Blake. What is a project, exactly?'

'Do I look like a walking dictionary?'

'No. You're the CEO of a start-up that's gambling everything on this one project of yours. So: how would you define a project?'

'A group of people doing something new,' Blake guessed.

'A little vague, but not bad,' Riley said. 'Here's Marguerite's take. *There is only one reason for a project: to create something both essential and exceptional.*'

'Sounds obvious.'

'Obvious once it's spelled out to you. And I believe your project reflects Marguerite's maxim. But we both know there are plenty of projects out there that reflect the vanity of their instigators.'

'Increasing the human lifespan matters deeply to most people,' Blake said. 'So our project's essential. And we're going to change the way the world uses DNA sequencing. So it's also exceptional.'

'And yet,' Riley reminded him, '*85% of projects fail.*'

'They crash and burn,' Blake mused.

Riley nodded. 'So how do you ensure Med•evolv falls into the 15% category, rather than the 85%?'

'Well, that's where my knowledge and experience both fall short,' Blake admitted.

'Here's your dilemma in a nutshell,' Riley said. 'The more you worry about failure, the more you rely on project management methodologies to keep your project on track. And yes, these methodologies do work like magic, but not in a good way. Like any skilled magician, they misdirect your attention.'

'Misdirect my attention? How?'

'By diverting you from the real cause of failure.' Riley took a deep breath. 'Here's what Marguerite has taught me, and what I've learned from every project I've been involved with. You can't achieve brilliance by following a rigid methodology. Projects are fluid, dynamic, unpredictable. If you want Med•evolv to succeed, you need to be fluid, dynamic, and yes, even unpredictable at times. Look for patterns. Look for risks. Look for opportunities. Think ahead—a long way ahead. Be the Grandmaster of personal genome sequencing. Don't be distracted by any sleight of hand.'

'Fluid, dynamic, unpredictable?' Blake smiled. 'Describes me to a T.'

'Except there's a special factor that makes project management so complex. Something you may struggle with.'

'What's that?'

'Perhaps Marguerite's third maxim will help you understand,' Riley suggested. *'**Processes neither make nor break a project. People do.**'*

Blake shuddered. 'You're right. Dealing with people is my Achilles heel. But despite everything you've said, surely you need some way to track what's going on. You always need some kind of project planning methodology.'

'Marguerite would never deny that. However, she would ask you this. What makes the greater contribution to your project—a piece of scheduling software, or the guy in the lab who's about to discover a new genetic marker for pancreatic cancer?'

Blake stared at Riley. 'Sweet baby Jesus...'

'Unlikely. But maybe if you pray for a miracle...'

'No, no, no. I've finally grasped your meaning. Taybridge is not managing his people. He's not looking ahead. He's just managing the methodology. Managing the process. And I'm not managing Taybridge.'

'Which means?'

'Well, I'm thinking of either the *Titanic* or the *Hindenburg*.'

'Nasty. So let's take some steps to avoid a disaster.' She passed him a pen and notebook. 'You might want to write this down.'

After leaving the café that Friday evening, Blake sent Taybridge a text. He scheduled a meeting for 8.00 a.m. the following Monday morning, in Blake's top floor office. Taybridge arrived five minutes early. Blake skipped the usual niceties, cutting straight to the chase.

'Thank you for your regular project reports, Aiden. They're most professional.'

'Of course, Mr. Stein.'

'However, I'm wondering if you'd be able to help me out with a minor problem.'

'In what way, Mr. Stein?'

'Come now, Aiden. I'd much prefer you call me Blake.'

'That wouldn't feel sufficiently professional, Mr. Stein.'

'Whatever you prefer, Aiden. The thing is, I believe this project carries some major risks. I'd like to know more about these risks, and how we're planning to address them.'

'Risk management is an essential component of project management,' Taybridge replied.

'And what is the biggest risk this project presents to you?'

'Not hitting our milestones. But at present, everything is on track.'

'So it seems,' Blake agreed. 'Let me quote from the report. "Identify correct Pantone colors for internal project stationery." Check. "Develop risk indices for mitochondrial genome sequencing." Check. "Roll out Cat 5 Ethernet cabling to Level 6 annex." Check.'

'Consistency of branding is important for any project, Mr. Stein. It can strongly affect staff morale. The risk indices you mention are indispensable. And the cabling—'

Blake cut him off. 'That's all penny-ante stuff! Of course you need Cat 5 cables. But that's not the essence of the project. Can you tell me what we are trying to achieve here?'

'Of course.' Taybridge sounded offended. 'We are developing a cheaper form of DNA sequencing which updates in real time, allowing us to build an international database linking to disease risk.'

'So how will you know if we've succeeded?'

'I regard a strike rate in the ninetieth percentile to be a substantial achievement.'

'You do realize the success of this project hinges upon us developing a breakthrough technology that includes mitochondrial DNA in the genome sequence?'

'Of course, Mr. Stein. Which is referenced in the risk indices you mentioned earlier.'

'So if you can tick off 95% of your objectives—Pantone colors, risk matrices, infrastructure rollouts and the like—but miss 5%—say, achieving accurate mitochondrial DNA sequencing—how would you categorize the project?'

'I would regard the project to have achieved a high standard of compliance with Prince2 methodologies.'

'Would you like to know how I'll be judging the success of the project?'

'If your suggestions can provide a basis for objective measurement, I would be most interested.'

'I want a product that blows everyone away. I want this to be the iPhone of real time human genome sequencing. I want the CEOs of our biggest biotech and insurance companies to come up to me and say, "Med•evolv has achieved something awesome, man. You rock!"'

Taybridge shook his head. 'Those criteria are not Prince2 compliant, Mr. Stein.'

$\mathcal{T}$WO

'I can't slip away right now, Blake, I have a critical meeting at 9.30. How about lunch?'

'I'm not sure I can wait that long.'

'You'll have to. Find something to help you while away the next couple of hours!'

After hanging up, Riley switched her phone to silent.

Toying idly with her latte, Riley listened to Blake's recap of his conversation with Taybridge before passing judgment. 'Sounds to me like the responsibility for project management is going to fall on your shoulders,' she said. 'But first, you have to neutralize Taybridge. Otherwise, he'll drive your project headlong into an iceberg.'

'True true. But his VC buddies have welded him into place. How do I handle that?'

'Easy. You made a pitch that excited them. They decided to back you. But $200 million's a decent sum. Somewhere along the way, they became nervous. Putting Taybridge into the project manager role is a symptom of that. And along the way, somehow you've lost status. Who's the VC angel who's holding the purse strings?'

'Julian Meadows. Harvard grad, West Coast native. His family made its first fortune selling victuals to the '49ers back in the gold rush, and diversified from there.'

'So there's a great-great-grandfather Meadows somewhere in the family tree who had an eye for the kill. What's Meadows like? His hard-bitten ancestor, or a soft preppy type?'

'He's sophisticated enough, but I see the fire in his eyes. When the lions bring down the wildebeest, he'll be in for his share.'

'And does he want to bring down the wildebeest himself?'

Blake thought for a moment. In his mind, he went back over his meetings with Meadows. He remembered a man who asked tough questions, who knew how to push his buttons. 'I believe he does,' he said at last.

'So. He's your project sponsor. He's paying the bill. Somehow, Taybridge won his confidence. You need to win it back.'

'And how do I do that?'

'Remember Marguerite?'

'*Processes neither make nor break a project. People do.*'

'That's her. And when I sent you off to test whether or not Taybridge was a maker or a breaker?'

'You gave me another dot point from Marguerite's PowerPoint. *A vivid, shared vision unites a team.*'

'Meaning?'

'People who share a clear vision, but it's more than just a managerial platitude. Marguerite meant people who feel so compelled to realize that shared vision they'll crawl across a desert with you to achieve it.'

'Exactly. So you're going to meet with Meadows, and align your stars with his.'

Fifteen minutes later, Blake and Riley had sketched out their plan on his tablet. He pushed his chair back, indicating his desire to leave.

'One thing before you rush back to the office,' Riley said.

'What's that?' Blake replied, glancing at the time on his iPhone.

'You haven't asked me how my meeting went.'

'Oh. So—um—how did your meeting go?'

'Just a tip, Blake. If you want people to take an interest in you, take a real interest in them.'

'I'm interested,' Blake replied. 'It's just that I'm keen to try out the ideas we just discussed.'

'Remember the old *quid pro quo*. It wouldn't hurt you to take an interest in your pseudo-twin sister, would it?'

'What choice do I have? Tell me about this meeting. I'm guessing it didn't quite work out the way you'd planned.'

'You're right on the money, Blake. New federal regulations have imposed a substantial compliance burden on Bank Pacific West. Six different departments are affected—and each one has its own ideas about the best response.'

Blake tried—unsuccessfully—to stifle a yawn.

'Boring, isn't it?' Riley said. 'But the money that funded Pyrouette and the money that funds Med•evolv doesn't just fall out of the sky. It's channeled through the financial services sector. So this affects you more than you'd like to admit.'

'You're probably right. I'll do my best to listen. But first—more coffee!'

Once Blake had returned from the counter, Riley began her tale. 'The proposition I took to this morning's meeting seemed obvious enough to me. Find the similarities between the reporting requirements the new federal laws impose on each department, and build the system from there. Automate the reporting process by integrating it with the existing system. Trouble is, we don't have one IT system, but nine. And they refuse to talk with each other.'

'But you only have six departments!' Blake exclaimed.

'Sure. Two of the departments share the one system. Another department has a legacy system no one these days can decode. And the remaining three departments run a patchwork quilt made up of the remaining seven systems. Talk about a nightmare. To make matters worse, each department head has proposed a quick and dirty fix that will only entrench the problem deeper in the long run.'

'So what did you propose?'

'Simple. Develop one modern system that covers all their reporting needs, migrate the data from the existing systems, and roll the new system out across all six departments.'

'Sounds slow and clean, rather than quick and dirty,' Blake quipped.

'Clean, yes. Slow, no. We can meet the federal deadlines if each department head commits the resources to my team. But no, they can't see any further into the future than the end of next week. So they knocked back my proposal.'

'Wow. So what are you going to do now?'

'Another quick tip, Blake. Before you ask me what I'm going to do to solve the problem, ask me how I feel about it.'

'Why would I do that?' Blake asked. 'Not trying to be smart—I just don't know the answer.'

'Because right now I'm exceedingly pissed off, and I need to vent.'

'And I need to—ah—listen and stuff?'

'Listen and empathize. So sit back, and don't gulp down that mocha too quickly.'

Ten minutes later, Blake slouched back in his seat and rubbed his hands over his chin. 'I never knew you were such an expert in medieval forms of punishment!' he said. 'Or that you hated the Director of Business Lending so much! Do you feel any better now?'

'Much better. Of course, I'd appreciate it if you'd keep what I just said confidential.'

'Of course. So it's called a shrew's fiddle, is it?'

'Actually, it's a rhetorical device called black humor. Helps me relieve the stress of working with such backward thinkers. When I'm interacting with them I need to remain completely professional, so by listening you've allowed me to really blow off some steam.'

'It's been a pleasure. So—what will you do now?'

'Simple. I'll gather the data I need to prove my case. I'd hoped it would be intuitively clear to these fools, but it wasn't. Time to step up and show them how to save some real money. That'll clinch the deal—but it means spending a week or two on something I consider superfluous.'

'Does that annoy you?'

'It's part of the game. This is important, Blake. Just about everything humans do, especially in business, is nothing more than a game. But when everyone wears suits, they start taking themselves too seriously.'

'I never wear a suit,' Blake boasted.

'So I've noticed. But sometimes, if you want to play the game right, you'd better dress for the part.'

Meadows caught Blake off guard. Instead of meeting at an upscale restaurant, he suggested a little Italian seafood place down by the water. 'They do the best chowder outside of Boston,' he declared. 'Meet you there at midday Thursday.'

So Blake returned his Italian suit to its hanger, and went with his usual garb: khaki chinos and a button-down shirt. And more than a dash of self-doubt.

'What sort of message is Meadows sending by meeting somewhere cheap?' he wondered. 'That he doesn't value my work?'

Blake pushed his plate of chowder to one side. His verdict? Capable enough, but hardly brilliant. He kept this thought to himself. He hadn't come to evaluate the meal.

'I have a question for you, Mr. Meadows,' he said. 'About personalities.'

'Call me Julian,' Meadows said, smiling. 'And your question is?'

'It's not easy for me to ask.'

'And yet, it's an important issue for you.'

'It is. Every day, I concentrate on achieving the best outcome for our project, rather than on the personalities of my team members. But sometimes, it seems to me that these different personalities can stop the team reaching peak performance.'

Meadows nodded. The stillness in his eyes encouraged Blake to continue.

'I find it hard to talk with Taybridge. It's like we're from different planets.'

'Perhaps you are. What do you want him to hear?'

Blake hesitated. Meadows' question turned him inside out. What was he trying to say?

'That there are some issues that worry me about the project.'

'Issues?'

'Well, problems, I guess.'

'So how might you have broached this subject in the past?'

'I'd have said, "There are some problems we need to discuss."'

'Does Taybridge need to discuss them?'

'I believe so.'

'Has he ever raised these problems with you?'

'Never. Which is why I need to discuss them with him.'

'Ah. So you need to discuss them. He doesn't.'

'Well, it would be in his best interest.'

'And when you've tried this tack in the past, how has Taybridge responded?'

'Sarcasm. Avoidance. Passive aggression.'

'Are you surprised?'

'I was. But now I figure I must be doing something wrong. I'm just not sure where I'm making my mistake.'

'Back up a little. In the past, you'd tell him there were problems that needed to be discussed. What tone of voice would you use? What body language?'

'Hard to say,' Blake mumbled.

'Really?' Meadows smiled. 'You mean no one ever told you?'

'People said I was pretty hostile. I'd puff out my chest. Raise my voice. Curse.'

'Do you know why?'

Blake frowned. Surely Meadows didn't need him to spell out the obvious. 'Because I was angry.'

'Were you really angry?'

'Angry? I was livid with rage!'

'And what was beneath your anger?'

Blake felt like the coyote, the moment after he runs off a cliff when he looks down and realizes the only thing holding him up is his own wishful thinking. 'There's nothing beneath my anger,' he replied. 'Just a long terrifying drop, followed by rapid deceleration when I hit the ground.'

'Trust me, Blake. There's always something beneath the anger. Once you work it out, we'll continue this conversation.'

Blake excused himself, and went to the restroom. With his fingers dancing on the screen of his iPhone, he tapped out a rapid-fire conversation with Riley.

Quick! What's beneath
my anger?

You for real?

Don't rile me, Riley!

Can't you work it
out for yourself?

Wouldn't be asking
you if I could...

Same as anyone's
anger. Fear.

Fear? What do you mean?

Who are you angry
with now?

Taybridge.

Because he won't do
what you want?

True true.

You're frightened of
losing control.

Of Taybridge?

> Exactly. But you never really
> controlled him, did you?

Blake returned from the restroom a little wiser, but not as wise as he hoped to appear.

'I worked it out,' he told Meadows. 'I'm angry because I can't control Taybridge. Deep down, I'm not angry. I'm scared of failure.'

'Did you just text a friend?'

'I have to understand what's going on. I can't work it all out by myself.'

'So imagine you're having a conversation with Taybridge—just the two of you—and it turns bad. You can't always text someone for the answer.'

'How did you know what I was doing?'

'I might be fifty-two,' Meadows said. 'That doesn't mean I'm clueless.'

Meadows sipped his mineral water and watched silently as Blake rehearsed the conversation with Taybridge in his mind. Blake approached the exercise like a chess Grandmaster: advancing a line of argument, then retreating when he sensed a trap. His mind moved quickly. Before Meadows had finished his drink, Blake had realized the limitations of his strategy. He tapped his finger lightly on the table.

'OK,' Blake declared. 'I've worked out several things.'

'I'm delighted to hear that.'

'First thing. There's no point in me blaming Taybridge for my problems.'

'Correct.'

'Second—you had good reasons for appointing Taybridge as project manager.'

'Also correct.'

'Third—I don't know what those reasons are.'

'Our reasons will be revealed in good time.'

'Fourth—you must also have good reasons for appointing me CEO.'

Meadows looked askance. 'Are you doubting yourself?' he asked.

'Not really,' Blake mumbled. 'It's just...' He took a deep breath. 'Right now, I'm not sure what I'm adding to the project.'

'That hardly fills me with confidence, Blake,' Meadows said. 'Med•evolv is your baby. You conceived it, you have a clear vision of the product you're building, and the genius needed to make it happen. Provided you work closely with Taybridge.'

'Which brings me to Point Five,' Blake said. 'If I want to work with Taybridge, I need him to listen to what I have to say. But he won't listen to me if I don't listen to him.'

'The Golden Rule, revisited.'

'How do you mean?'

'Listen to others as you would have them listen to you.'

Blake took a moment to reflect on Meadows' words. 'Interesting,' he said. 'So you're telling me I need to rethink my whole relationship with Taybridge if I want to make it work?'

'Re*think*?' Meadows added some acid to his voice.

Blake did not quite catch the meaning in Meadows' tone. 'Absolutely. So far, I've been an abject failure. So I need to start from scratch.'

'Does it only involve thinking?'

'Mr. Meadows—Julian—I'm a geek. Everything happens up here.' He tapped the side of his head. 'Three pounds of gray jelly. It's all in the mind.'

'Including your anger and fear?'

Blake stared at Meadows. He wanted to speak but both his brain and mouth came up empty.

$\mathscr{T}$HREE

Blake toyed listlessly with a sugar sachet. 'I totally lost control of the conversation,' he griped. 'Remember how I was going to align my stars with Meadows'? Didn't happen. I started whining about Taybridge, and Meadows turned everything back on me.'

'That's great news!' Riley replied.

'Great news? What are you on about?'

'Any time I complained to Marguerite, she always hit me with the same reply. *Who's holding the buck?*'

'Must have become rather monotonous,' Blake shot back.

Riley sighed. 'You've heard the expression, *the buck stops here?*'

'Of course. In this case, it stops with Meadows. He's the money man.'

Riley shook her head. 'That's not what it means. And that's certainly not what Marguerite meant, either.'

'So why don't you enlighten me?'

Riley reached out, plucked the sugar sachet from Blake's fingers, and placed it on the other side of the table, well beyond his reach. 'Picture this. It's the 1830s. You're in a log cabin on the banks of the Missouri, playing poker with three other ruffians. They don't trust you any more than you trust them. The only illumination in the room is the dim light from an oil lamp. You know they'll try to cheat you if they can.'

'I don't know the first thing about poker,' Blake grumbled.

'Then it's time you learned,' Riley said. 'Keep your eye on the prize, Blake. You pull out a hunting knife, and drive the point into the table in front of you. It marks you as the dealer. After you've dealt, you pass it to the next man. If he refuses to deal, he passes it around the table.'

'I still don't know why I'm in Missouri, of all places.' Blake shuddered.

'The handle of your hunting knife is carved from buck horn,' Riley explained. 'It's called a *buck*. When you say *the buck stops here*, it means you are responsible. You're holding the knife. You deal.'

'Are you saying project management is the same as playing poker?'

'You're overcomplicating this, Blake. Marguerite would say yes, in some respects project management is like playing poker. But what did you say as you were turning that sugar sachet over and over in your hand?'

'That I had lost control of the conversation with Meadows.'

'Exactly. So who's holding the buck?'

Blake screwed up his face. 'Meadows, obviously.'

'Did he lose control of the conversation?'

'No. I did.'

'So who's responsibility is that?'

Blake stared at Riley. 'Will you pass me some sugar?'

Blake tore the sachet with his teeth, and poured the contents into his mocha. The sugar floated briefly on the foam, before sinking below the surface. He picked up his teaspoon, and slowly stirred it in.

'It takes me a while to absorb all this,' he confessed. 'With Meadows, I felt like a little kid talking to an adult. I balked from the start. I didn't want him to think I was stabbing Taybridge in the back.'

'Very wise,' Riley affirmed. 'If you did, you'd have lost most of your credibility in an instant.'

'So I took a circular route. I thought I'd reach a point where we could talk about my vision for the project. But I let Meadows derail me.'

'All of which tells me you handed Meadows the buck, right from the start.'

'Lamentably true. But you seem to be implying I'm responsible, not Meadows.'

'I am. After all, you passed the buck to him.'

Blake let the silence consume him. He stared down at his mocha. 'I think I'd prefer a long black,' he said. 'This one's too sweet.'

'Let me order this time,' Riley said. 'Like you said, you need time to make sense of all that's happened.' She stood, and then paused. 'There have been many times at Bank Pacific West when things haven't gone my way. Marguerite says that whenever this happens, I have to take responsibility for my actions—and my reactions.'

Riley was glad to see a small queue in front of her at the counter. 'More time for Blake to ruminate,' she told herself. She knew how he felt. The first time Marguerite had pulled that *who's holding the buck* business on her, she'd wanted to scream.

To be fair, it did not seem like Riley's fault. A freshly minted lawyer joined her team. Jennifer's role was central to the project—to interpret the Financial Accountability Act, and ensure the procedures the project team developed complied with the law.

From the start, Riley had some doubts about Jennifer. Not about the keenness of her legal mind—Jennifer could slice and dice the law with the aplomb of a skilled sushi chef. No, it was Jennifer's apparent lack of assertion that kept Riley awake at night. Could she stand up for herself when she knew she was right? Sadly, Riley's instincts proved correct. Jennifer let the experienced bankers on the project team push back against her interpretation of the law—with almost fatal results for the project.

Riley thought she had kept her eyes on the prize. Until the morning when she was summoned to meet the bank's Chief Legal Counsel, Milton Reynard. Reynard was not a man given to small talk. As soon as Riley entered his office, he slammed her draft proposal down on his desk.

'If we adopt the procedures you're suggesting, the Directors are staring down the barrel of a five-year jail term. And financial penalties of up to $500,000 each. I thought you had a lawyer on your team precisely to manage this risk!'

Riley froze. She had no idea how to respond. She did not even understand the cause of the problem. 'So you're saying...'

'That this proposal is not sufficiently robust to protect the bank's Directors from the full force of this legislation. It's as though it's been written by junior banking staff who want to protect their bonuses, without realizing the larger implications of their actions!'

Riley returned to the table, where Blake still seemed stunned. 'Let me explain what happened to me,' she offered. Blake nodded. So Riley recounted her confrontation with Reynard.

'Of course, I met up with Marguerite for drinks that afternoon. She ordered a dirty martini, but just the one. "Always keep your wits about you, young Riley," she counseled. Which is amusing, because I'm only a few months her junior. I hoped she'd let me off the hook—that it was all Jennifer's fault for being weak, or the bankers' fault for being greedy.

'Of course, I was mistaken. Marguerite led with a jab. "Who's holding the buck?" she demanded.

'I mumbled something foolish. Marguerite closed in to finish me off with a barrage of shots to the head. "You're the project manager. You knew Jennifer couldn't speak up for herself. You knew the bankers would want to preserve their entitlements, regardless of the consequences. So: who's holding the buck?"'

Blake felt a sudden urge to defend Riley. 'You can't possibly be across every last bit of detail in the project, surely?'

'Don't try to excuse me, Blake. It sounds like you're trying to excuse yourself. I needed to make sure the bank and its Directors were protected, and I had failed. Luckily, I still had time to redraft the proposal, but we ended up over time and over budget. I hardly covered myself in glory. Marguerite was right: I was supposed to be holding the buck, but I'd passed it down the line.'

'So what should you have done?'

'Simple. I should have checked in with Jennifer to see if she needed more support before she met with the bankers who were drafting the procedures. I should have explained the big picture to those bankers, so they understood the possible consequences of their decisions. And I should have reviewed the draft with Jennifer, to ensure it protected the bank's interests.'

The waitress arrived with Blake's long black. He took a deep satisfying sip before continuing the conversation.

'While you were ordering the coffee,' he began, 'I put two and two together—somewhat reluctantly, I might add. You're saying it's my fault I let Meadows run away with the conversation. Which means I need to be stronger next time I speak with him.' Blake paused, and looked around the coffee shop. It was packed with people engrossed in their conversations. Everyone seemed so confident, so excited by their own brilliance. 'Deep down, I guess I'm frightened I'm not strong enough.'

Riley reached across the table, and placed her hand on top of Blake's. 'Until now, I'd have said you were right. But you've just convinced me otherwise.'

'How?' Blake demanded. 'By sounding weak?'

Riley shook her head. 'Not weak, but honest and vulnerable. Everyone has doubts. Some folk are just better at masking them with bravado or anger. By the way, it's not your fault you let Meadows dominate the conversation. I'd rather say you were responsible for the result.'

'What's the difference?' Blake seemed genuinely perplexed.

'Assuming you're at fault is a form of self-flagellation. Taking responsibility means accepting the consequences of a mistake, and moving on. I'd rather move on than wallow in self-pity.'

'Fair enough. I'd like to move on too. So how do I take responsibility for the outcome next time I see Meadows?'

'I think he's already given you a clue,' Riley suggested.

'Must have blinked,' Blake replied, 'because I certainly missed it!'

'We don't hold the buck because it's the logical thing to do,' Riley said. 'We hold it because we've mastered our anger and our fear. Here's what I suggest you do...'

After Riley had outlined her plan, she glanced at her phone. 'I'm late,' she said. 'Sorry, but I have to rush.'

'Before you go,' Blake said, 'there's something I need to ask.'

'Better make it quick,' Riley said, gathering up her phone and keys.

'Last time we met, you had this issue with the Director of Business Lending, among others. How did it all turn out?'

Riley's smile came from the corner of her mouth. 'I'm glad you remembered, Blake. I'm more relaxed about it now. There's still some data I need to collate, and then I'll be ready to present to the Board. If I can, I'd like to run things past you first. You know, to gain a male perspective?'

'I'd be glad to.'

'Next time, then. I really have to run.'

Blake left the coffee shop with two contradictory emotions swirling about in his mind. On the one hand, the dull ache of depression. In the past he had let his anger and fear dominate his decision making. How could he ever change? Surely his behavior was the result of a genetic predisposition, something he could never shift? On the other hand, he felt a keen sense of hope. He did not need to remain mired in the past. Riley had convinced him he could change, and she was seldom wrong. If she could believe in him, then he owed it to her to repay her trust. He could choose to believe in himself.

Usually when Blake felt down he would seek out the company of strangers. A lively bar where he could drink, proclaim his genius, and lift his spirits. Going out kept his depression at bay. This time, however, he knew it would be wiser to follow Riley's suggestions. He dropped into a stationers, and bought a notebook. It fitted comfortably into his hand, and he ran his fingers over the Celtic knot pattern embossed on the antique gold cover. Although brand new, it felt centuries old. It reassured him that everything was going to turn out alright.

When Blake returned to his apartment, he poured himself a mineral water, rather than his customary scotch. 'Always keep your wits about you, young Blake,' he told himself. He opened his notebook, and at the top of the first page he scrawled a heading: *WHAT AM I AFRAID OF*?

He paused for a moment, and ran his fingers over the cream-colored acid-free paper. Why would he defile this elegant notebook with his neurotic ramblings? Then he remembered Riley's command: 'Buy the most expensive notebook you can find. Use it to work through your feelings. Don't worry about the cost. You're worth it.'

Blake stopped chewing the end of his pencil, and began jotting down every fear that came to his mind.

'Failure.

'Ridicule.

'Not knowing. I'm afraid of not knowing everything...'

$\mathcal{F}$OUR

'This is a pleasant surprise,' Meadows said. He and Blake had just taken two of the best seats in the restaurant at the Park Regis. 'The lobster here is exceptional. And as you've offered to pay, I may as well indulge myself.'

'I'd be disappointed if you took anything less than the best,' Blake replied. 'It's my way of thanking you.'

'For a cheap bowl of chowder? No thanks are necessary.'

Blake smiled. 'Not the chowder, Julian, but the conversation. To be more specific, your willingness to push me out of my comfort zone.' Meadows opened his mouth to speak, but Blake stayed him with his hand. 'Before you say anything, yes, I did consult an old friend. Her name's Riley. I need someone I trust implicitly as a sounding board.'

'I'm glad you did,' Meadows said. 'If Riley can make you more human, then more power to her.'

'We're working on that. I realize that if I'm going to have any significant input into the Med•evolv project, I need to work better with others.'

Meadows sipped from his glass of mineral water. 'The project won't succeed without you,' he said. 'From the start, you convinced me with the strength and clarity of your vision for the company. But you can't make it all happen by yourself.'

'True true. In the past, I've held on too tightly to all the elements of a project. I thought that was what I needed to do. But I just ended up disempowering my people. I squeezed all the enthusiasm out of them. I'm not going to make that mistake again.'

'How are you going to stop yourself?' Meadows asked.

'Good question. Last time we met, you asked me about my anger and my fear. I've worked on both of them. I've realized I'm a perfectionist, that in the past I've driven people to meet my standards, without considering their needs. That has to change.'

'So how will you change it?' Meadows gave Blake a searching look. 'Talk's cheap. The proof is in your behavior—and the outcomes you achieve.'

'Already underway!' Blake proclaimed. 'I've listed all the things I'm afraid of, then I've gone back and identified the key events in my life that shaped my fears. I filled twenty pages of my journal last night. When I sit back and contemplate them all, I see how easily my fears turned into habits. There's this one memory that's particularly vivid—'

This time, Meadows raised his hand. 'Blake, I'm not your confessor. I don't really need to hear about your self-analysis. There's a simple reason for this. I'm deeply invested in Med•evolv. I want it—and I want you—to succeed. But I can't be your therapist.' Meadows plucked a business card from his wallet, and slid it across the table to Blake. 'If you need to talk with someone who can help you, ring this number.'

Blake glanced at the card. 'I don't need a psychologist!' he exclaimed.

'I think of her more as a high performance business coach,' Meadows said.

'You can recommend her?' Blake asked.

'I've referred several people to her,' Meadows said. 'Geniuses all. In each case, her intervention meant the difference between success and failure.'

'I meant in your personal experience,' Blake insisted.

'Her contact details are on the card,' Meadows replied, more gently than Blake deserved.

Blake picked up the card by its corners. The idea of seeing a psychologist both repelled and thrilled him. 'Thanks,' he said. 'I'll give this the consideration it deserves.'

Meadows' lobster arrived, along with Blake's roasted vegetable stack. The lobster looked spectacular, and Blake briefly regretted his choice. As soon as he lifted his first mouthful of flame-grilled aubergine and red pepper to his lips, he remembered why he had chosen a less lavish meal.

'In my experience, lobster always overpromises and underdelivers,' he told himself. 'Trust your first instincts, Blake!'

He and Meadows ate in silence. Once they had both finished, and the plates had been cleared from the table, Blake resumed their conversation.

'I want to level with you,' he said. 'I've been having some difficulties with Taybridge. I'd like to know what's negotiable, and what's not.'

'Is your problem only with Taybridge?' Meadows asked.

Blake had not anticipated this question. 'Well, yes! I mean, everything else is going smoothly...'

'There are no problems within the team?'

'Not as far as I'm aware.'

'Tell me, Blake: how would you describe your leadership style?'

Blake hesitated. Another question that wrested control of the conversation from his grasp. 'Well,' he began, 'I see myself as a can-do sort of guy.'

'Excellent!' Meadows blessed him with a radiant smile. 'So what have you done?'

Blake frowned. Up to this point in the project he had been busy—extremely busy—but what had he actually achieved? He reviewed the last six months, and realized he had nothing concrete to report.

'Well, it's still early days,' he said, hoping he didn't sound too defensive.

'Taybridge has accomplished a lot over the last six months,' Meadows said. 'Have you seen his reports?'

Blake decided not to chance his luck further. 'Maybe I've come at this project from the wrong angle,' he admitted. 'I thought it was all about having a big idea, then riding people until they brought it to fruition. I mean, the history of business is replete with stories about visionary bosses who drive their team to the point of exhaustion, but still develop a product or service that revolutionizes the world.'

'True,' Meadows replied. 'At the same time, there are many stories of failure that don't become part of the mythology. Sure, the spectacular failures are noted, like the Edsel and the Apple Lisa. They're excused because Ford and Apple achieved so much with their iconic products. However, you don't hear about the everyday failures—the ones that quietly lose their backers significant sums of money.

'I'm a venture capitalist. I prefer to back winners. Med•evolv has the potential to change the way we manage human health, both individually and collectively. Neither of us wants it to fail. That's why I gave you the name of someone I trust.' Meadows glanced down for a moment. 'Someone who's helped me in the past.' He fixed Blake with his eyes. 'You're on the verge of a massive personal shift. Or a massive personal crisis. And unless you shift, Med•evolv is guaranteed to fail.'

The waiter returned with a dessert menu. 'Just a pot of Darjeeling,' Meadows said. 'Blake?'

'You're having tea?' Blake asked, his voice subdued.

'I am. I find it aids my digestion—much more so than coffee.'

'Then I'll try the same. Two pots of the Darjeeling, please.'

'Of course, sir.' The waiter departed.

'So you're well aware of the risks involved in this project,' Blake said.

'I am,' Meadows replied. 'Which is why I wanted Taybridge to set up the initial processes.'

'Why is that?'

'Two reasons. Because he is meticulous. So I can be sure he will attend to every last detail while we're setting up the project. I couldn't trust you to do this. You'd leap right into the technical challenges, without getting the fundamentals right.'

'Like the Pantone colors on the logotype?'

'A trivial example, but yes. There's a second reason I wanted Taybridge involved.'

'And that is?'

'I knew he'd clash with you. Your personalities exist at opposite ends of the spectrum. Risk minimizer versus risk maximizer. I need you both working together if the project is to succeed.'

'But you seem to relish the fact that Taybridge and I don't get on!'

'*Relish* is the wrong word, Blake. From where I'm sitting, I see Taybridge performing well. On the other hand, I've been less impressed with your contribution. I'm glad you contacted me last week. If you hadn't, I'd have called a meeting—and we'd be having a much harder conversation than the one we're having now.'

Blake took only one key message from Meadows' comments: *Taybridge is performing well, while you are not*. He began to tremble inside.

'We both want Med•evolv to succeed,' Meadows added. 'Neither of us wants another Pyrouette.'

'Do you think I'm going to crash and burn a second time?'

'I'm more interested in whether *you* think you'll crash and burn again.'

It took all of Blake's nerve to resist the urge to rearrange the silverware again. 'The thought has crossed my mind,' he whispered.

'I'd suggest viewing Pyrouette as a successful apprenticeship,' Meadows said. 'Sure, you attracted a lot of negative press. But you also learned what not to do. You won't make the same mistakes twice.'

'You're probably glad you never invested in Pyrouette,' Blake said.

'I cannot make money without taking risks. I see Med•evolv as a risk worth taking. You're keen to redeem yourself. Throw yourself into another project, and hope like hell it works.'

Blake chose his words deliberately. 'I'm guessing you're not quite as optimistic as I am.'

Meadows smiled. 'I'm immensely optimistic about your ability to imagine a brilliant product, one with the potential to transform the world. Your ability to manage a start-up? I'm not so sure. From my observations, you believe everyone on the project team should be highly gifted. Am I correct?'

'That wouldn't hurt,' Blake conceded.

'And yet, the reality is that while your team has many gifted people on board, none of them are the same as you. They all have their own strengths and weaknesses. As a project leader, it's your role to get the best out of them, while minimizing the risks posed by their weaknesses.'

'They're good people,' Blake said. 'It's not as bad as you suggest.'

A slight frown creased Meadows' forehead. 'I'm not suggesting anything negative. I'm simply being realistic.'

Blake steepled his fingers. 'I just proved my perfectionism again, didn't I?'

'You did. But let's decide to make this work. How can I support you? And what do you need from Taybridge?'

'To ensure the project succeeds?'

'Exactly.'

Blake pushed back in his chair, and ran his hand over his lips. 'I need you to be totally honest with me. Tell me when I screw something up. Meet with me regularly so we can talk things through.'

'I can do that. My group has over $100 million riding on your success. We hold a majority stake in Med•evolv. I've personally vouched for you to three other VCs. We'll spend all the time you need. And Taybridge?'

'That's a harder question. He has done all the groundwork you mentioned, and I'd have missed a lot of it myself. I guess he and I need to work out who is responsible for what.'

'Last time we met you said he and you seemed to be from different planets. How are you going to make it work?'

'Good question.' Blake thought for a moment. 'When you asked me if there were any problems on my team, I couldn't give you a straight answer. Perhaps I just don't look closely enough. Taybridge and I need to work together to identify any problems, and address them.'

'So, think about Taybridge. He's uncomfortable with risk. Anything else?'

'Well, he's good at managing detail. At the same time, I sense that he's into saving face. That he doesn't want to be seen to fail.'

Just then, the waiter arrived with the tea. Meadows poured out two cups and offered Blake the tiny jug of milk. Blake accepted, adding just a dash to his Darjeeling.

'And how would you describe me?' Meadows continued.

'That's easy!' Blake laughed. 'You're a big picture, connect-the-dots sort of guy. At first, you seem a little austere, but when people get to know you, you loosen up. I'd say it's more important to you for the project to succeed, than it is to avoid failure.'

Meadows smiled. 'Good calls. So how did you get me on board?'

'I told you a story about a girl called Gina. She knows her family has a history of breast cancer, so she's worried about her future. Gina has the Med•evolv base station installed in her bathroom. Once a week, she places a saliva sample in the base station, which undertakes a gene sequencing process over the Net, using the otherwise untapped power of distributed computing. The Med•evolv software expedites the process by isolating changes to Gina's established genome sequence. If mutations to her mitochondrial DNA indicate a heightened cancer risk, we inform Gina immediately. The health insurance industry subsidizes the cost to Gina. They're playing a long game, building up a much stronger risk profile as more people come on board with Med•evolv over the next twenty years. The aim is to reduce the cost of medical interventions by identifying cancers early, and also helping genetic researchers develop gene therapies which will eventually eliminate the risk of cancer.'

'And how did I respond when you told me about Gina?'

'Your eyes lit up. You connected all the dots. In fact, if I remember correctly, that last bit about gene therapies was actually your idea.'

'It was. How would Taybridge respond to the same pitch?'

'I doubt it would work. He's a different sort of person.'

'I agree. So here's your challenge. Think about Taybridge's personality. How do you bring him on board, so he can work with you to achieve a common vision?'

$\mathcal{F}$IVE

'Wow!' Riley steepled her fingers, and pressed them against her upper lip. 'Meadows has set you a real challenge.' She paused, the thoughts rushing through her mind at warp speed. 'But if I'm going to help you, you're going to need to trust me completely.'

'I know. Which is why we're meeting here'—Blake swept his hand around the lounge room of his inner city apartment—'rather than The Creamery. What we're going to discuss has to stay in this room.'

'Your inner sanctum!' Riley replied. She smiled inwardly. Despite all the money that had poured Blake's way from the Pyrouette deal, he still furnished his million-dollar apartment as if it were a student dorm. Not that she could criticize. Most of her furniture had arrived in flat-packs, too.

'Here's my problem in a nutshell,' Blake began. 'I can describe my project to Meadows, and he connects the dots. Connects them so well he coughs up over $100 million. And if I'm talking to some of my best technical people, they grasp the idea immediately. But Taybridge? I may as well be speaking Swahili. He just doesn't get it!'

'So you can't explain your project in a way that resonates for him?'

Blake gave Riley a searching look. 'Actually, I thought it was because he's a little slow on the uptake.'

'You mean he's less intelligent than you?'

'That sounds...' Blake searched for the right word, but he couldn't find one that didn't make him sound full of himself.

'Condescending?' Riley suggested. 'Judgmental? Arrogant?'

'Stop that!' Blake demanded. 'At one level, I know he's not stupid. And yet, at another...' Blake scratched the side of his face. 'It's so obvious what I'm trying to achieve. Why can't he see it?'

Riley thought for a moment. 'Tell me what he does well,' she suggested.

'That's easy!' Blake replied. 'All the detailed work we need to do to set the project up. You know, all the stuff I'm not really that good at doing.'

'I know, Blake.' Riley smiled. 'Drove us nuts when we were working on that robotics contest. You'd be debugging a program you'd flashed to an Arduino board, and then you'd break off to run some Cat 5 cable across to network in a new printer. Halfway through you'd realize we'd run out of breadboards, so you'd log on to Mouser to place a rush order. You drove us to the point of planning your disappearance, you realize!'

Blake shook his head, and bit down on his knuckles. 'Thanks for not covering my head with a spray jacket and forcing me into a white van at knifepoint,' he said.

'I thought I had to keep on top of everything, but it was more than I could manage.'

'You didn't trust the rest of us enough to delegate any of the work,' Riley said. 'Even checking our inventory of breadboards!'

'But while I was futzing about, you were sitting quietly in the corner, writing the code that connected our sensors to the Arduino board. That was the core of the project, and you managed it without any fuss.'

'Except for the fuss you created by panicking at the slightest provocation,' Riley pointed out. 'And yet, I remember several flashes of brilliance that made the difference between success and failure. Flashes that seemed to erupt from your mind like lighting.'

Her praise failed to lift Blake's mood. 'You see my dilemma?' he pleaded. 'My mind works best when I'm trying to stop everything from collapsing around me. My most compelling ideas only emerge in the middle of a crisis. How can I work effectively with someone as anal as Taybridge?'

'It's a challenge for you,' Riley admitted. 'But your co-workers might prefer a lower chaos-to-brilliance ratio.'

'We won second place in the robotics contest.' A note of defensiveness crept into Blake's voice. 'And we did sell our IP for a handy sum of money.'

'True,' Riley conceded. 'But it didn't work so well at Pyrouette, did it?'

Blake dared not answer Riley's last question. 'No,' he thought to himself, 'Pyrouette was an unmitigated disaster. And the responsibility rests entirely with me.'

He stood up, crossed to the fridge, and poured two glasses of soda. He had slurped down almost half of his before returning to his seat, and handing the second glass to Riley. 'We only achieved a decent result with our robotics project because our team was so small. Five students. Mind you, three of them couldn't wait to abandon ship—all the pain I caused was more than they could bear. And it cost me, too. I stressed myself out, ended up with a stomach ulcer. But I didn't learn much from our success.'

'It takes an exceptional person to learn from success,' Riley said. 'Most of us only learn from failure—and then, the lesson usually has to be repeated many times before we take notice. But let's return to Taybridge. What does he need from you?'

'What a weird question,' Blake said. 'Shouldn't it be the other way around?'

'No doubt there's much you need from him. Which is another topic. Right now, it's more important for you to consider what Taybridge might need from you.'

Reluctantly, Blake pondered Riley's question. 'Well, before I can answer you, I need to consider his strengths. Hard as it is for me to admit, he's good at imposing a sense of order on the project. He's good at achieving results when the task is straightforward, and even if it's complex.' Blake searched around for the right words. 'What I mean is, he's not going to generate new ideas, but he's able to manage complex tasks, and bring them to fruition.'

'So he's good at working with the known, rather than the unknown?' Riley asked.

'That's it!' Blake said. 'So I guess he needs me to utilize his expertise, rather than treating him as if he's an unimaginative drudge.'

'Have you been doing the latter?' Riley asked.

Blake nodded slowly. 'Afraid so. As much as I hate to admit it, I've always treated him with just the slightest smear of contempt. He's no fool; he's picked up on it. So now our interactions are pretty chilly.'

'So he needs you to treat him with respect, and to let him bring his skills to bear?'

'I guess so.'

Riley chose her next question carefully. 'Is he really an unimaginative drudge?'

Blake looked at her with an expression of pure shock. 'Isn't it obvious? I mean, it might be a harsh term, but doesn't it contain a kernel of truth?'

'Evidence, please!' Riley demanded.

'Well, he just...' Blake scratched his head. 'After all, he...' Blake frowned. 'Come on, Riley! I've not seen him do anything the slightest bit creative!'

'Have you looked?'

'Well, no...'

'Blake.' Riley lowered her voice. 'Here's the thing. In my experience, everyone—even those people we think of as dull—is capable of coming up with new ideas.' She paused, and then corrected herself. 'Not just capable— in most cases they want to improve things. They want to share their ideas. They want to be heard. Taybridge has done all the groundwork to set up your project. How do you know he hasn't introduced a whole raft of incremental improvements into Med•evolv's business systems?'

'Fair enough,' Blake conceded. 'But incremental improvements don't mean as much as...'

'As much as what? As much as your flashes of genius?' There was a sarcastic edge to Riley's voice. Blake shivered a little inside.

'I guess that sounds rather uppity,' he admitted.

'If you want to work with Taybridge, you'll need to get over yourself,' Riley said. 'I can't do that for you. It has to be all your own doing. But I can help you plan your conversation with him. By the way, what's his first name?'

'I think it's Alan,' Blake muttered. 'No, perhaps it's Aiden...'

'Probably best to be sure about that,' Riley said. 'Now, you're going to tell me as much as I need to know about Med•evolv to help you dig your way out of your current hole. And once we're done, I'd like to run through my presentation for Littler, our Director of Business Lending.'

'To give you that all-important male perspective?'

'That's what I'm after.'

'Well, you give me your female take on Taybridge, and I'll be glad to help you out.'

'So tell me, Aiden,' Blake said, 'is this all your own work?'

Taybridge watched Blake from the other side of the boardroom table. Some of his initial wariness had dissipated. 'Of course not,' he said. 'I'm hardly an expert in IT or HR. I did develop the Gantt Charts, the resourcing profile and the risk register. But I outsourced the specialist work to the relevant experts.'

Blake flicked through the papers on his desk. 'As a result, we have a computer network with a uniform operating system. All the programs work seamlessly with each other—and with our R&D lab. From my perspective, this is a minor miracle—I've always worked with a kludge of computer systems that refuse to play nice. By establishing a shortlist of preferred suppliers, you've shaved 15% off our procurement budget. And you've worked with recruitment to bring me a strong mix of new grads and experienced scientists. I didn't realize we expected so much from you.'

'Some of this work has taken me beyond my remit,' Taybridge confessed. 'But we only have a small team. All these things had to be done.'

Blake rubbed his hands over his chin. 'I never thought I'd be saying this to you, but I'm impressed. Most of these items would have fallen through the cracks if you hadn't caught them.'

For the first time, Taybridge allowed himself a faint smile. 'I'd never thought you'd be interested in this,' he said. 'I assumed it was too low-level for you.'

Blake smiled. 'I'd thought the same, I must admit. But I'd like us to work more closely on this project. We can't do that if we don't understand what we both bring to the table.'

Taybridge nodded. 'My role is to support Med•evolv. To remove any barriers to the company's success. End of story.'

'I guess I took all these project underpinnings for granted,' Blake said. 'So tell me about this.' He pushed a slim booklet across the table to Taybridge. '*Prototype Guidelines: From Laboratory to Manufacture.*'

'I heard a consultant speak about this subject at a conference last year,' Taybridge said, 'and the need to ease the transition from research to production stuck in my mind.'

'He wrote it for you?'

'She. Professor McFadden had twenty years' aerospace experience before moving to academia. And this is no off-the-shelf product. She's tailored it to the needs of biotech and IT.'

'No need to defend your choice, Aiden. I just assumed—my bad.'

'It's not too soon, from your point of view?' Taybridge asked.

'The Board still needs to endorse this document, correct?'

'Absolutely. This is just a draft I'm sending around for consultation.'

'It addresses a problem I've experienced before,' Blake admitted. 'You develop a prototype, but without thinking about the costs and complexity involved in manufacture. This sets out some clear guidelines that will hopefully stop us making this mistake again.'

'You don't mind?' Taybridge asked. 'I was worried I might have overstepped the mark. You might feel they were too restrictive.'

Blake leaned back in his chair. 'I know what you're saying,' he said. 'I must confess my initial reaction was far from positive. I don't like anyone imposing any limits on my work.

'But then, I realized that limits are exactly what I need. When I was in college, I came across this eccentric lecturer in my coding class. He had a whole rant he liked to give about the value of limits. Guess what he used to illustrate his point?'

'I have no idea, Blake,' Taybridge replied.

'Shakespeare! He talked about Shakespeare's sonnets, and all the rules Shakespeare had to follow when writing one. Imagine telling that to a lecture theater full of eighteen-year-old IT students. But then he said something that made me sit up and take notice. He said, "A sonnet is just an arcane form of coding. A form of coding that has given us some of the most powerful love poetry in the world. When I look around the room, I'm guessing many of you would like a little more romance in your lives. You want romance? Learn to crack the code."

'Then he brought us back to reality. "The restrictions the sonnet form imposed allowed Shakespeare's genius to flourish. Someone in this room could be the next Shakespeare of coding. Someone who's not overawed by the restrictions of their craft—but rather, liberated by them."'

Taybridge allowed Blake's comments time to sink in. 'So, you're telling me you're happy with these guidelines?' he asked.

'They'll help unify the project team, and keep us on track,' Blake said. 'You've done well.'

'My team and I have done well,' Taybridge corrected him. 'Everyone has contributed. I couldn't have done it alone.'

Blake felt chastened. 'Fair call,' he admitted.

'Very few of these ideas are mine,' Taybridge said. 'They came from everywhere. I just pulled them together. I found that even those people who seem really black and white in their thinking could generate great ideas. And that people who seemed chaotically creative were able to impose a sense of order from time to time. I see the two as complementary, rather than oppositional.'

'When you talk about chaotically creative, were you referring to me?'

Taybridge blushed. 'It's not my place to judge,' he said.

'But you've hit on a key issue,' Blake replied. 'I thrive on chaos. Or rather, I believe I thrive on chaos. That element of uncertainty helps me see connections no one else has seen before. I'm worried that working in such a well-organized project will inhibit my ability to make those connections.'

Taybridge glanced down at his feet. 'I've been worried about that too,' he admitted, 'although from a different perspective.'

'Tell me more,' Blake said.

'Well, everything I've shown you today is under control. But I'm not stupid. I can see there are some major advances you need to make before Gina can manage her health from the comfort of her bathroom.'

'You're referring to the current realities of gene sequencing technology?' Blake asked.

'Exactly. At present, Gina needs to send her saliva sample to a laboratory packed full of complex and expensive equipment. How do you miniaturize those machines to fit next to her electric toothbrush?'

Blake laughed out loud. 'I'm glad you're not a venture capitalist, Aiden! Because you'd be asking me questions I might struggle to answer.'

'This question has been keeping me awake at night,' Taybridge confessed. 'I don't know how you're going to solve these problems. So I've focused on the things I can control, rather than those things I can't.' He hesitated. 'These little bits of detail I've attended to? I could foresee them because I have enough experience to know what will happen if they're overlooked. It's not the things I can foresee that worry me. It's the things I can't see, the things I can't anticipate. I'm hoping you'll be taking care of the future for us.'

'I'll do the best I can,' Blake promised. Then he remembered Riley exhorting him to be the Grandmaster of personal genome sequencing. A shiver of doubt ran down his spine. 'But maybe...'

'Maybe?' Taybridge repeated.

'Maybe there are some things I can't anticipate,' Blake said. 'Even so, I feel confident we'll succeed. History tells me that over the last fifty years we've miniaturized and digitized many processes, which we once would have thought impossible to do. I'll be working with our R&D team to identify all the markers that might indicate a genetic mutation, and using software to zero in on these markers. How we'll solve all our problems, I don't yet know. But I know one thing for certain. Once we have unwrapped some of the secrets of our DNA, we'll keep going. And I'm hoping the more secrets we unwrap, the easier our task will become.'

Taybridge heard a note of doubt in Blake's voice, but lacked the courage to challenge him. Instead, he asked, 'Have I made the project too antiseptic for you?'

'Perhaps. Which brings me to another item on my list of discussion points. I need to build a sandpit.'

'A sandpit? I know nothing of best practice when it comes to sandpits,' Taybridge said with the faintest twinkle in his voice.

'I mean I need to create my own chaotic, exploratory world. But I'll wall it off, so it doesn't affect the rest of the project.'

'I'm intrigued,' Taybridge said. 'Can I help you with that in any way?'

'I think I can create a mess without any help!' Blake replied. 'But if you're interested, I'll show you how it evolves.'

'I am interested,' Taybridge confessed. 'The better I understand how your mind works, the better I can assist you.'

Blake leaned back in his chair, and stretched his arms out. 'There's one more agenda item I'd like to put on the table,' he said. 'We're not going to resolve everything today. It's going to be an ongoing conversation for the life of the project. But we do need some clarity about our respective roles.'

Taybridge folded his arms across his chest. 'Well, I've assumed I'm responsible for logistics. Making sure everything is in place for all the members of the team, whenever they need them.' He paused. 'But the Board also wants me to track the progress of the project.'

'Hence those traffic light reports that lob on my desk each Friday.'

'Your tone of voice suggests they don't really delight you,' Taybridge said.

'You're right,' Blake replied. 'I just find the layout patronizing. It's a bit too much like kindergarten for my liking.'

'But that's what the Board members seem to want.' A note of confusion crept into Taybridge's voice. 'I mean, that's the established protocol for project management reporting.'

'Oh,' Blake said. 'So it's not that you think I'm particularly slow on the uptake?'

Taybridge shook his head. 'It's a standard template that comes with our project management software.'

'So we're kind of locked into using it?'

'Not really. I can change it if you like.'

Blake raised his hand. 'Let me talk with the Board members first. Maybe they're happy with it.' He hesitated. 'From my point of view, I'd rather you and I sat down to speak frankly about any problems that arise.'

'Like we've done this morning?'

'Yes. And this may sound strange, coming from Doctor Anarchy, but I think we'd be wise to minute the outcomes. In case either of us falls under the proverbial bus.'

'Sure,' Taybridge agreed. 'That way our successors will be able to bring themselves up to date.'

'So you're responsible for project logistics, and project reporting,' Blake said. 'You and I are responsible for discussing any problems that might arise. And you'll be recording our decisions somewhere on your project management software.'

'That sounds like a clear summation,' Taybridge said. 'And you?'

'And me?' Blake smiled. 'I guess I can go and play in my sandbox.'

Taybridge shook his head. 'I can't agree,' he said, his voice firm. 'My job title is Project Manager. You might be Med•evolv's CEO, but you're also our project's real leader. You have the vision that inspires our project team to drag themselves out of bed each morning, even when they've been working into the small hours the previous night. They need to know your eyes are on the prize, that you're looking after the future for them. I can't inspire that kind of loyalty. And it means you can't just go play in your sandpit. Not unless you invite a dozen coders, a dozen engineers and a handful of geneticists to join you.'

Blake took a deep breath. 'You're asking me for a commitment I've so far failed to meet.' He shook his head. 'This is what you need from me. It's what Meadows and his fellow angels need; it's what the Board needs; it's what our coders and engineers and geneticists need too. You know why I've failed to step up to the plate?'

'No idea,' Taybridge admitted. 'It's something I've been asking myself every day for the last six months.'

'Because I'm shy.'

'Nonsense!' Taybridge exclaimed. 'I've seen you down at the bar, a glass in hand, telling all and sundry how brilliant you are. People hang on your every word!'

'That's just bluster,' Blake confessed. 'There are so many ways this project could fail, and everyone seems to be looking to me for the answers.'

'No.' Taybridge shook his head. 'They're looking to you for inspiration. While you remain confident, the rest of your team remains confident. But if you succumb to doubts...'

'This conversation has to remain in this room,' Blake said.

'It will,' Taybridge assured him. 'My job is to support you in every way possible. My silence is part of that support.' He hesitated. 'I also need to report our progress to the Board. But everything you've told me today is privileged.'

Taybridge stood up and shook Blake's hand. He turned to leave, and then paused. 'So how did your Shakespearian sonnets turn out?' he asked.

'Better than I expected,' Blake admitted. 'I wrote a couple, and used them to woo a young English major. We had some good times together before the relationship ran its course. A few years ago, I bumped into her downtown. She told me she had kept my sonnets, along with all her student assignments.

'"Well," I said, "those sonnets must not have been as bad as I'd feared."

'She just chuckled at me. "Blake," she said, "they were the worst sonnets I'd ever read. Mawkish, clumsy, sentimental. But I appreciate the fact that you tried."'

$\mathscr{S}$IX

'I really owe you, Riley!' Blake said. He had ordered their coffees at the counter, and returned to their table with more enthusiasm than he'd experienced for a long time. 'Aiden and I really hit it off. And it's all down to the plan we devised. I followed every step to the letter.

'First, I set aside all the prejudices I'd formed since first meeting Aiden. It was like you said, *Don't fight the prejudice. Just accept that you may not know the full story.*

'Second, I asked open-ended questions. I told him the project seemed to be going really well, so I was curious to learn if he'd encountered any obstacles. And if he had, how did he overcome them?

'Third, I listened. Normally, I'm not that good a listener, because I've got too much to say myself. But once I'd established a curious mindset, I found myself fascinated by everything Aiden told me.

'Fourth, I acknowledged the work he's done. He surprised me. I thought he was just another accountant type, the kind of guy who always says *no* to everything. But I'm starting to think he could actually be an asset.

'Fifth, I used the rapport we'd developed to expand the conversation, and clarify our different responsibilities. In fact, it turned out he was as eager as I was to nail down our roles. Maybe he and I will be able to work together effectively in the future.'

'So it's that simple, is it?' Riley snapped. 'One quick chat and he's your best bud? You want a productive working relationship with Taybridge, you're going to have to work on it.'

Blake felt her words punch him right in the stomach. He looked at her and for the first time noticed the anger in her eyes. 'What gives?' he asked.

'Just that everything in my world has turned to crap,' Riley said.

'I don't understand. Last time we met, you were flying high.'

'Remember how Littler asked me to prepare some data to justify my case?'

'Littler is your Director of Business Lending?'

'One and the same. He finally came back to me. You know what he said?'

'I'm guessing it wasn't positive.'

'He said, "Your data is inconclusive." And when I tried to pin him down on the detail, he made some pitiful excuse, and ran away.'

'But you ran through that presentation with me. It was flawless. Your findings were indisputable!'

'So I thought. Littler says otherwise.'

'I have no idea what you should do next,' Blake admitted.

Their coffees arrived. Riley took a sip from her long black. 'But I do,' she said. 'Once you've finished your mocha, we're taking a drive.'

'A drive? Where to?'

'Across the bridge. Marguerite has a home office overlooking the harbor. She'll know what to do.'

Their Uber driver pulled up in front of Marguerite's house: three tiers of concrete and glass cascading down the hill. Sprays of bougainvillea softened the concrete slabs. Blake hesitated before stepping out of the car.

'What's the protocol here?' he asked Riley. 'I mean, Marguerite has attained mythic status in my mind. What if I say or do something that makes me look stupid?'

A smile spread across Riley's face. 'You're feeling nervous!' she exclaimed. 'That's kind of cute...'

'Are you going to help me out here,' Blake grumbled, 'or are you just going to practice your Cheshire cat smile?'

'In the interest of allowing our driver to continue with her day, may I simply suggest you act respectfully? You've nothing to fear. Marguerite always said there's no question too stupid to be asked.' She paused. 'At least, not for the first time. She expects you to pay attention. She doesn't like to repeat herself.'

'Nice,' Blake muttered.

'You know what he told me? "The data is inconclusive." The illiterate jerk! He doesn't even know the difference between data and datum! What makes him think he can critique my data sets?'

Blake sat, a little warily, on Marguerite's settee, with his back to the view. To his left, Riley was busy working up a fine head of steam. She waved her arms wildly, and there was so much anger in her voice she almost choked on her words.

Marguerite smiled at Riley's impotent rage. A slender young woman with close-cropped gray hair and multiple piercings, she conveyed a sense of deep self-assurance. 'May I see your data?' she asked.

Riley passed her tablet across to Marguerite, who scanned the data quickly. Blake watched as she navigated around the screen, moving her fingers with the precision of a croupier. Her eyes never deviated from the tablet. She reminded Blake of a hawk: remorseless, sharp, and unrelenting.

After a minute or two, Marguerite placed the iPad back on the table.

'Tell me, Riley,' she said, 'can you give me one good reason why Littler wants you to fail?'

'Fail?' Riley's mouth hung open, gaping like a goldfish. 'I don't understand.'

'You've clearly presented your case. Your data back you 100%. There can only be one conclusion: do what you recommend. Which tells me Littler is playing you. So: what's in it for him?'

Riley shook her head. 'Beats me,' she replied.

'I've yet to see anyone beat you,' Marguerite said. 'Take a walk. Clear your head. Come back with an answer. I want to talk with your friend for a while.'

Blake felt a sense of panic quicken in his stomach. 'My God!' he thought. 'She's engineered the whole thing. Leaving me alone with this woman who's going to ask me questions I have no way of answering.' He felt the fear rising to his throat as Riley left the room.

'Blake,' Marguerite said, skipping any small talk, 'there's one thing I need to know about your project. How will you know if you have succeeded?'

'Well, I'll be able to buy myself a Maserati,' Blake joked.

Marguerite smiled. Not a smile of amusement; more a smile that promised some future pain for Blake. 'Because a Maserati is going to transform the way we practice medicine in the West?'

Blake swallowed hard. He wanted to backtrack, but feared losing face.

'You seem uncomfortable,' Marguerite observed. 'If I've taught Riley anything, it's the importance of building a vivid, shared vision of your project outcomes. Has she shared that with you?'

Blake nodded, but his mouth stayed mute.

'She would also have explained how important it is to align that vision with your sponsor's vision. Have you done that?'

At last, Blake found his voice. 'I tried to. I met with Julian Meadows—he's our biggest investor—intending to have that conversation. But I sort of found myself sidetracked.'

'So you didn't take charge of the conversation?'

'Well, there was a rather delicate political issue I needed to raise...'

'Let me guess. You thought you'd try to sneak round the back, rather than knocking on the front door.'

'Well, maybe...'

'How much capital have you raised?'

'Two hundred million.'

'And your projected rate of return over ten years?'

'Four hundred percent. That's conservative.'

'You've identified your project's scope?'

'Extensively. Thousands of pages in the report.'

'And yet you can't tell me if you and Meadows are on the same page. That's not a question, by the way. It's a statement of fact.'

Blake resisted the temptation to stare at the floor. He lifted his head, and looked Marguerite directly in the eye. 'Sounds like I've stuffed up,' he admitted.

'Correct. So let's talk about unstuffing things. Are you the right person to do this?'

'How do you mean?'

'You cannot lead a project if you're easily intimidated. You need to work with other people. Sometimes you need to inspire them. Sometimes you need to confront them. But you also need to understand the politics of project management. All projects are political. You can't avoid that. It's your job to step up and unite your team. Lead them to victory. You up for that?'

'Well, I'm more of a geek than a Napoleon,' Blake admitted. 'My expertise is in coding software.'

'You good at that?'

'I'm brilliant.' Blake refused to lie.

'What's stronger?' Marguerite challenged him. 'Your ownership of the project, or the outcome?'

'Aren't they the same thing?' Blake asked.

'No.' Marguerite's tone of voice told him she would brook no argument. 'The project is the sum of everything your team members do. It's easy for them to get caught up in their petty squabbles if their elements of the project begin to fail. The outcome, in contrast, is your holy grail. It's the tantalizing prospect that persuades your people to focus on the end result, rather than why Bob was given a cubicle with a window. So, which is stronger? Your ownership of the project? Or your ownership of the outcome?'

'The outcome,' Blake declared.

'Good.' Marguerite's voice softened. 'Now all we have to do is avoid you becoming another Steve Jobs.'

Blake hesitated before replying. 'Isn't Steve Jobs a great American hero?'

'Of course. But consider this. He took over the Macintosh project and drove it to completion because he had a vision of how personal computing could be transformed. Not because he was a brilliant coder. He achieved his goal. A year later, he was forced to resign from Apple. He was thirty years old. How old are you?'

'Twenty-eight.'

'Do you want to spend twelve years in the wilderness?'

'No. I want to lead Med•evolv for the next twenty years.'

'Excellent!' Marguerite exclaimed. 'So here's what we're going to do.'

Riley returned an hour later. To her astonishment, Marguerite's lounge room was covered with large sheets of paper. The sheets of paper were covered with what appeared to be random scribblings. But in Riley's eyes, they formed a coherent plan.

'Looks like you two have been busy,' she remarked.

Blake looked up. Riley recognized the enthusiasm she saw burning in his eyes. 'There's so much I've done wrong. But Marguerite believes we can remedy it.'

'We?' Riley asked.

'Marguerite and I are planning a series of meetings with the project team. They'll help us regain a sense of purpose—something I've allowed to slip over the last few months. But before it kicks off, I need to have another conversation with Meadows.'

'That's wonderful,' Riley said. 'And my problem with Littler?'

Marguerite set aside her marker pen. 'Blake and I are just about finished here. Take a seat. What I'm going to say concerns you both.'

Riley collated several sheets of paper before sitting down. She placed them to one side on the floor. 'I think I know what Littler wants,' she said.

'Good work,' Marguerite replied. 'Please continue.'

'The proposed government regulations have the potential to severely damage our business lending sector,' Riley explained. 'Littler does not trust us to develop a set of procedures that limit this damage. He's been burned in the past by overly zealous bureaucratic types. Personally, I see this new legislation as an opportunity for us to grow the business lending sector—but I need Littler on board!'

'So how will you manage this challenge?' Marguerite asked.

'By meeting with Littler and working through the opportunities this presents. Opportunities which are contained in the data he has not yet read.'

'You know this because?'

'If he'd read the data he would be leaping out of his seat and embracing me in his excitement. He needs them explained to him.'

'And before you meet with Littler?'

'I'll touch base with the project sponsor, to make sure he has my back. Although I don't like to deploy the heavy artillery, it's good to know it's there.'

'All projects are political,' Marguerite said.

'Every last one,' Riley replied.

$\mathscr{S}$EVEN

'Thank you all for joining us for these two days.' Marguerite stood in front of a wall of smartboards. The hum of the air-conditioner muffled the noise of the traffic passing outside the conference room. 'As your invitation stated, we're seizing the opportunity to look at the focus of this project. Not because you're off course, but because these conversations are necessary when you're working on a project evolving as rapidly as yours.'

The project team members nodded. Predominantly under thirty, a mixture of Caucasian and African and Asian descent, maybe three-quarters male. IQs off the chart, keen minds at the ready. Each one held a tablet and a stylus, linked by wi-fi to the smartboards that lined the room. The technology transformed the front room of a shabby Spanish Mission style motel into a collaborative workspace. Cheap enough to not trouble the budget; far enough out of town to demand an overnight stay; sufficiently run-down to create a slightly seedy ambience: the El Paso had everything Blake and Marguerite required of a conference venue. They had already relieved their guests of their smartphones as they entered the room: it's hard to commit to vision planning when you're distracted by a tweet, or the sweet addiction of Candy Crush.

'First up: you don't know me. My name is Marguerite Lavigne. I help organizations manage projects. Here's the simple truth: it's best you manage your projects yourselves. But 85% of projects fail. You know why? Because you're all brilliant individuals. But projects don't succeed on individual brilliance. They succeed on *collective* brilliance. This is where most projects screw up: they don't harness the collective brilliance of the team.

'So I'm here to set you up, and then get out of your way, so you can all go back to what you do best: transforming the way we deal with the genetic hand we've been dealt. Disrupting the old paradigms of health insurance. Using emergent technology to monitor our health status in real time. And as soon as I step out that door at the end of our sessions, I'll go and help other organizations change their part of the world.

'Some quick background. My last project was classified. I can't name the country I worked for; I can't name the government department involved. Let's just say it was far larger and more complex than this project. We succeeded, although none of you will ever know the details. However, your chances of being alive in five years' time have risen exponentially because of our work.'

An excited hum of speculation filled the room. Marguerite raised her hand. 'The methods we used on that project are the methods I'm offering you over the next two days. As you know, Julian Meadows is your angel investor. He's here this morning to share some of his vision with you. Blake Stein and I have spent some time with Julian, and we're excited by the possibilities his support offers this team. Julian can't be with us for two whole days, and we need some time to clarify key aspects of the project. However, Julian will return this afternoon, to check in and hopefully sign off on some key initiatives.

'We'll be using those tablets to share our ideas with each other. But technology is not a panacea. What matters most is not what we capture electronically. It's what we capture up here.' Marguerite tapped the side of her head. 'Luckily, I know you all have plenty of room for big ideas.' She paused while the team members chuckled at her comment. 'Does anyone have any questions before we begin?'

For a few moments, no one stirred. Marguerite allowed the silence to build. Eventually, a young man at the back of the room raised his hand. 'Why do we need two whole days for this?' he demanded, his voice trembling. 'We all have deadlines to meet. When we get back to the office we'll be two full days behind.' His question was greeted with a smattering of applause—maybe one-third of those attending shared his views.

Marguerite knew he had not heard anything she said during her preamble. She also knew better than to attack him, or put him down. Her smile conveyed a sense of serenity. She'd heard similar objections before, and they no longer troubled her. 'Thank you for being so honest—I didn't catch your name?'

'Thomas,' the young man whispered.

'Thomas, it sounds like quite a few of your colleagues feel the same way—am I right? Can I have a show of hands?' Marguerite cast her eyes around the room. A sizable minority raised their hands in the air. 'Again,' she said, 'thank you for your honesty. Would any of you be able to tell me how the project is tracking?'

Wilson Clay, one of the crack coders, spoke up. 'I know *I'm* on track,' he asserted, 'but as for the electronic engineers...'

'Sounds like this is an issue for you, Wilson,' Marguerite said. 'Would any of the electronic engineers like to respond?'

Thomas raised his hand again. 'Well, maybe if you could write firmware that actually interfaced with our chips...'

Marguerite stepped forward, and raised both hands in the air. 'Let's not declare open warfare,' she suggested. 'At least, not yet. It sounds to me as though there may be some issues that haven't fully surfaced. Am I correct?'

Wilson and Thomas spoke in unison. 'Absolutely!' they both said.

Taybridge glared at them. 'Why haven't you told me about this?' he asked.

Wilson shrugged his shoulders. 'You never asked. Besides, the scoping plan said we'd be using Version 0.93 of the firmware for initial testing.'

Marguerite sensed a growing discontent in the room. 'Gentlemen and ladies!' she announced. 'Before we get down into the trenches and join in hand-to-hand combat, let's step back and take in a bird's-eye view of the battlefield. Here's the plan for today and tomorrow.

'First, we're going to build a shared vision of success for the project. How will it look, how will it sound, how will it feel?

'Second, we're going to break into work teams, and work out what you need to do to bring your vision to life. And you're going to share your discoveries with the other teams, because as Thomas and Wilson have just shown, you guys are relying on each other to ensure Med•evolv's success.

'Third, we're going to develop some options, before deciding on the best plan to take this project forward. Options that you'll present to Julian Meadows later tonight.

'Tomorrow, we'll start to nail down the details of your chosen approach. Who does what, with whom, and how long it will take. By late tomorrow evening you'll have identified all the key issues, and you'll have begun developing your final project plan. And yes, when you go back to work on Wednesday, you'll have a two-day backlog to clear. But let me promise you this. If you don't invest the time now, at some stage down the track you'll encounter problems that will take you months to solve. So where would you rather invest your energies?'

This time, a thoughtful silence filled the room. The threat of rebellion had faded. Marguerite looked into the faces of the participants, and smiled again. 'Your project is challenging because it's cutting edge. It's poised on the cusp where science and computing and social transformation meet.' She turned and drew three overlapping circles on the smartboard, and labeled them before turning to her audience once again. 'What you achieve over the next two days and the next two years has the power to change the world beyond recognition.

Please welcome Julian Meadows, your VC archangel, who's here to share his vision for Med●evolv!'

The room erupted into applause as Meadows stepped up to the lectern.

The next two days passed in a flash. Blake watched in amazement as Marguerite guided his team through a series of verbal and intellectual gymnastics. As she led the sessions, Blake jotted down the questions she asked. But he didn't fully understand the cognitive leaps she made. How she would link a *sotto voce* remark one day to an anonymous comment appearing on the brainstorming page the following afternoon. How she could hold so many ideas in her head and reflect them back accurately to the group. 'If she thinks I have the talent to do this without her, she's sorely mistaken,' he thought.

After Meadows finished his speech, he apologized for leaving—he had a critical meeting to attend in the city—but promised to return that evening to review their progress. With Meadows gone, Marguerite was free to explore the team's uncensored reaction to his vision. Gradually, she helped them develop a common understanding of success. How it would feel. How it would look. How it would sound. Even how it would taste. Blake agreed to lay up some bottles of Krug Vintage for the launch party. He could almost feel those oaken notes dancing on his palate.

Just as everyone's imagination moved into overdrive, Marguerite stepped in, and cooled their excitement. 'Of course, a launch is just a launch,' she said, fixing each individual in turn with her steely eyes. 'And you did not join this team because you want to share a magnum of fine champagne. You joined because you want to transform healthcare. You joined because you want to save lives. The Krug is just a symbol of success. It's not your end goal.

'And neither is your launch function. The launch is just another milestone along the road to achieving your end goal: bringing a highly evolved personal genome sequencer to the market.' She paused, and smiled at the group. 'Unless you achieve your real goal, the Krug will disappoint you. Your happiness will last as long as it takes for the bubbles to evaporate from your tongues.'

Her comments recalibrated the mood within the room. It became more somber, more focused. Deftly, Marguerite helped the team build a consensus around success. Then, she broke the team up into subgroups, to explore a range of questions around roles and resourcing. 'What do you need to do to make your vision come to life? What do you need to do to create that highly evolved personal genome sequencer?' Once the subgroups had completed that assignment, she came back with some tougher questions. 'What resources do you need? How long do you need to complete your critical chunks of the project? And what do you need from each other?'

During this exercise Taybridge sat in with the other subgroups with only one question in mind: *How can I help you achieve your goals?*

In the hours before Meadows' return, all the pieces of the puzzle fell into place. Marguerite made it all seem effortless, although Blake knew he and his colleagues were straining every sinew to achieve the result they desired. In the end, they had three plans to present to Meadows. One resembled Blake's dream car, a Maserati: fast, spectacular, and expensive. Another resembled a Bentley sedan: not quite as fast, more substantial, and almost as expensive. The third resembled a mid-range BMW four-wheel drive: dependable, versatile, and less expensive than the other two.

'It's between the Bentley and the Beemer,' Blake told himself. To his disappointment, the team declined his suggestion of automotive codenames. Instead, they used relatively neutral colors: Red for the Maserati, Black for the Bentley, and Blue for the BMW.

Meadows' return that afternoon electrified the team. Some had doubted his commitment; he proved them wrong. Looking drained after a long day of meetings with key VCs, Meadows rolled his sleeves up and stepped right into the middle of the debate. A couple of team members became so impassioned they overstepped the mark, expressing their views in ways which Blake found disrespectful. He glanced across at Marguerite. She seemed content to let the argument run, although she kept a weather eye on Meadows' mood.

Indeed, the session ran over time. Marguerite sent a runner to the kitchen asking them to push back dinner for an hour. Eventually, Meadows settled back in his chair. 'There are elements of Red, Black and Blue that we need to incorporate. Marguerite, have we captured them all?'

'We have,' she affirmed.

'And we need to stay within the Black Plan's budget,' he added. 'I'll be staying tonight, so I can either work with you to rough out the plan, or disappear for a while if that helps you more.'

As the group broke up, Meadows walked across to Wilson, who had been unable to restrain himself during the discussion. Some of his comments on the plan had descended into *ad hominem* attacks on his colleagues.

'Blake tells me you're a very talented coder,' he began. Wilson nodded. 'Just a hint,' Meadows continued, 'next time you're involved in a discussion like this, I'd suggest you play the ball rather than playing the man. Doesn't matter who it is. Senior investor or an intern. That's the best way to ensure your career prospers.'

He patted Wilson on the shoulder. 'You're welcome to join Thomas and me for dinner tonight,' he suggested. 'There's a particularly rare Barossa Shiraz on the wine list I'm keen to try.'

After dinner, Blake and Marguerite sat at the bar, soaking up the retro-chic ambience of an unrestored 1960s motel. Blake ordered a spiced mojito, while Marguerite settled for her usual—a dirty martini. Together, they worked through the core issues that had emerged during the day. Different team members flitted in and out of the conversation. Some added value; others seemed intent on idle chatter. Marguerite welcomed them all.

'Hope you're noting the contributions each person made this evening,' she said to Blake, after the last team member had stumbled off to bed.

'What am I looking for?' Blake asked.

'You're assessing their value to the project,' Marguerite explained. 'Every time a team member speaks with you, he or she is staking a claim, making a request. Some are saying, *I have something valuable to contribute*. Others are saying, *Please notice me. Please* like *me*. Which is more valuable to you?'

'The former, obviously.'

'And what will you do with the latter?'

Blake shook his head. 'They seem to need more attention. I'm no shrink, but maybe something went awry with their toilet training.'

Marguerite laughed. 'This is important, Blake. ***Processes neither make nor break a project. People do.*** Taybridge has recruited a group of talented high achievers for you. Does this make them all equal?'

'I guess not—but I expect them all to perform to the highest possible standard.'

'Then you're going to be disappointed. It doesn't matter how talented people are. Every group forms its own bell curve. Here's what you'll find.

'A small cadre of team members gives you 100% commitment and 100% effort. These people are gold. Treat them as such.

'A much larger group ends up putting in about 80%. You can stretch them for a short period of time with tight deadlines and all-nighters, but they'll always fall back to 80%. Some managers hate this. They expect everyone to offer 110%. I say, *80% is better than 70% or 60%*.

'Take a moment to let that thought sink in. The people you have on the project are the people you have. You might want a more talented or committed team. But Med•evolv went out into a competitive employment market and chose the best people it could find. Sure, you'll need to keep recruiting as the project progresses, and sometimes you'll discover an absolute gem. But most people are going to give you 80%. That's the reality. You'll need to learn to live with that.

'You have another small group of plodders and grifters. The plodders offer up poor work, but think they're doing something worthwhile. The grifters know their work is not up to standard. They spend their time trying to disguise this fact. If they put the same energy into actually doing their job, they'd be up with your one hundred percenters.

'Some of this sub-eighty group are what environmentalists call a protected species. They have some level of political clout that prevents you from sacking them. Perhaps they're Meadows' god-kiddies. Perhaps they have photos of you snorting coke at the Christmas party. These folk are an absolute drag on the project—but for whatever reason, you seem to be stuck with them.'

'I can think of a couple who fit that description,' Blake said. 'Except for the coke. I'm completely clean.'

'Glad to hear it,' Marguerite said. 'Finally, you have some different views when it comes to buy-in. At the end of the retreat, you'll have a shared vision for Med•evolv. Some people will accept it without any struggle, but without the intense level of enthusiasm you bring to the project. It's not that they don't care. It's not that they don't believe in Med•evolv. It's just that they're not willing to die for you. They're your main group. Then there'll be a small group that refuses to accept the vision. Some of them will dress their opposition up in fancy language— *Just playing the Devil's Advocate, you know*. And a second small group will become evangelists for the project.

'Obviously, you'll value the one hundred percenter evangelists most highly. But eighty percenter evangelists are also useful. And if you find any apostate grifters, move them on as soon as you can.'

'Unless Meadows was present at their baptism,' Blake joked.

'I've seen you with Meadows. He likes you. So a family connection might not be enough to save someone who's undermining the project.' Marguerite paused. 'Mind you, if you sack your Devil's Advocate, another one's just as likely to pop up.'

Blake thought for a moment. 'Sounds a bit like the class clown, perhaps?'

'How do you mean?' Marguerite asked.

'When I was in school,' Blake explained, 'we always had a class clown. And he wasn't the same guy all the way through. If the class clown moved away, someone else always stepped in and took his place. Often, it was someone no one had previously thought of as a comedian. But it was like the class needed someone to fill the role. Just like these Devil's Advocates you mentioned.'

Marguerite gave Blake a look of approval. 'Quite possibly,' she conceded, 'you're wiser than you pretend to be.'

To the surprise of most team members—but not to Marguerite or Blake—Meadows stayed overnight, and joined the group for breakfast the next morning. He joined a group of coders at a table looking out across an empty parking lot.

'Are you leaving after breakfast?' the bravest of them asked.

'No,' Meadows replied. 'Wilson asked if I wanted to stay for the final day, to help map out the resources, deadline and accountabilities for the project. To tell the truth,' Meadows paused, and lowered his voice to a whisper, 'this is the most exciting part of any project. It's when you see everyone come to life. And to me, that's far more exciting than the meetings I had lined up for this morning.'

All the members of the team had taken their seats in the conference room with five minutes to spare. Marguerite began by reviewing the outcomes from the previous afternoon. She summarized them into a fourth and final plan, which she codenamed Green. On the top right-hand side of the smartboard she wrote a number—the budget for the project—and a second, much smaller number. This was the number of months allowed to bring the project to fruition.

'Julian and I are different,' she told the group. 'For me, the first day of a retreat is the most exciting. Watching a group form a shared vision of success never fails to fascinate me. But for Julian, the second day—when you take that vision and turn it into a series of personal commitments—is the most exhilarating.' She paused, and looked around the room, fixing each individual with her clear gray eyes. 'The fascination I felt yesterday turns to respect today. Because this is when each of you takes full ownership of this project.'

Marguerite broke the group up into functional work teams. Her instructions were simple:

1. Identify each task that needed to be done.
2. Estimate how long each task would take.
3. List the resources required to support it.
4. Identify any prerequisites and interdependencies.

'By *prerequisites*,' she explained, 'I mean any action item which needs to be completed before another action item can be commenced.

'By *interdependencies*, I mean any action item that requires input from more than one team. Just to be clear, any specific action item can be both a prerequisite and an interdependency.'

In addition, she appointed two or three people in each group as *emissaries*. As well as contributing their needs and commitments to the team plan, they were given an additional task: to flit between the different teams, and to report back on the prerequisites and interdependencies they saw arising. Marguerite chose well—she had identified those with the personality needed to quickly infiltrate another team's conversation, share ideas, and convey them accurately back to their home team.

For the entire morning the room buzzed like a beehive. Blake, Marguerite and Meadows circulated between the teams, deftly avoiding the emissaries as they rushed to and fro. Marguerite answered questions about the process, while Meadows answered questions about money and deadlines. Taybridge also sat in with the different teams, listening intently, identifying needs, and helping map out the issues.

Before long, Blake found himself engrossed in the programming team's conversation. He picked up a marker and began leading the conversation. Marguerite sidled up to him and tapped him on the shoulder.

'Blake, there's a question you might like to consider from the group over by the back wall,' she said. Blake glanced at her with a look that suggested he would comply within the next little while. Marguerite stared at him with a look that told him to comply immediately.

'You need to be democratic,' she whispered as she guided him across the room. 'You're leading the entire project, not just the coders. If the others think you're partisan, you'll lead the project to failure.'

After lunch, Marguerite kicked off what she described as her *Quilting Session*: bringing together the individual pieces to create a unified whole. The participants chuckled as she likened the task to a sewing bee. They stopped chuckling when she informed them that since the 1800s women had used quilting to express their political views, and to push for social reform—a nineteenth century disruptive paradigm, if you will. 'If you can work together as harmoniously and honestly as a group of Quaker women, then you are almost guaranteed of success. Their work was never trivial, and neither is yours.'

Gradually, a map of the project began to assemble itself on the smartboards lining the room. Although most of the tasks and timelines resolved themselves as the group discussed them, Marguerite marked some of them with a question mark. And while most of the prerequisites and interdependencies made themselves clear, a small number also attracted a question mark.

'Some of these issues may be insignificant,' Marguerite explained. 'Others may prove decisive. When you bring together a project team, you rely on the group memory to identify and address all the possible problems. These are the areas where our collective memory has failed us. Tomorrow, when you return to work, your first task is obvious: to eliminate all the question marks from your project.'

Wilson piped up from somewhere in the middle of the throng. 'Will you be there to help us?' he asked.

Marguerite shook her head. 'No,' she said. 'This is your responsibility. You'll either clarify the issues around each of these question marks, or identify them as risks to the project. We'll meet again in a few days' time to generate a critical path for the project. We cannot do that until you've addressed all the risks you've identified.'

'So is Blake going to lead us?' Wilson asked.

'I have a meeting with our lawyers tomorrow morning,' Blake said. 'There are some issues around IP that I need to resolve.'

'So who is going to keep us on track?' Wilson insisted.

'I think you'll find that's my job,' Taybridge said, with more than a hint of determination in his voice. 'After these two days, I know exactly what needs to be done to bring this project home.'

Wilson gave Taybridge a searching look. 'I'm glad somebody knows. Because we're going to need all the leadership you can offer.'

The team remained euphoric even as they boarded the bus. Meadows waited behind to congratulate Marguerite on her leadership of the retreat. 'You've drawn those three plans into one, and you've gained complete buy-in for our budget and our timeline. To me, that's a day well spent.'

'The team drew everything together,' Marguerite corrected him. 'I just guided them through the process.'

'Sure,' Meadows replied. 'But without your guidance, they'd not have achieved one-tenth of what they achieved over these two days.'

'I'll accept your comment with good grace,' Marguerite said with a smile. A black Lexus pulled into the forecourt. 'Is this your ride?'

'Yes. I'm looking forward to relaxing at home with a single malt scotch.'

'It's well deserved. I appreciate your support over the last two days. It has been invaluable—for me, and for the team.'

Blake walked with Meadows out to the car. As soon as he was out of Marguerite's earshot, he asked the question that had been puzzling him all day.

'Tell me,' he said to Meadows, 'how did you bring Wilson onside?'

$\mathcal{E}$IGHT

'How are your energy levels, Blake?' Marguerite asked as they watched Meadows' car drive off.

'Pretty depleted,' Blake confessed. 'Before we push on tonight, I could use a triple espresso!'

'I have a better idea,' Marguerite said. 'There's a Japanese garden two blocks down. The return walk is around forty-five minutes. Now's the time to reoxygenate our lungs, rather than recaffeinating our bloodstream!'

'OK,' Blake replied, somewhat reluctantly. Then he brightened up. 'I guess we can talk about the project along the way.'

Marguerite shook her head. 'That's exactly what we're not going to do!' she told him. 'There will be plenty of time to talk this evening. Right now it's important to exercise, breathe deeply, and clear our minds. Let the oxygen sharpen our minds, clear away all the fatigue we've accumulated over the last two days. I think you'll enjoy these gardens. They were planted by Zen monks. You'll catch glimpses of a koi pond framed by the branches of cedar and cherry trees, and you'll notice Shinto and Buddhist shrines half hidden among the foliage. Walk in silence, notice every step. We'll talk once we return.'

To Blake's surprise, he re-entered the motel with his mind completely revitalized. 'I must talk with Taybridge about establishing Japanese gardens on the roof of our building,' he told himself, only half joking. He returned to his room, showered and changed, and met Marguerite in the dining room for dinner.

'Before we do anything else,' she began, 'it's important to recap what we've achieved over these last two days. I'm going to spell out the key steps I've taken. The time will come when I'm no longer able to help you, so it's essential you know this.'

'But I couldn't possibly do what you do!' Blake protested. 'The way you draw people out, the way you weave everything together. If I tried, I'd fall flat on my face.'

'Do you see any bruises on my face, Blake?' Marguerite asked.

'Of course not!'

'Because they've healed. I've fallen on my face many times. Each time I picked myself up, and pledged not to make the same mistake again. The only thing I bruised was my pride. So before long I realized there are times when it's essential to put my ego to one side. If it doesn't serve the project, let it go.'

Marguerite pulled out her tablet and a stylus. 'I'm going to jot down the key points for you to take away. I'll airdrop the document to you once we're done.

'First, Maxim One: *85% of projects fail.* You remember this from Day One?'

'Absolutely. I wanted to argue with you. What's with the 85%?'

'There are two different figures,' Marguerite explained. 'Thirty years ago, it was 50%. Now, with all our advancements in project planning methodology, the failure rate is 85%. This means that leading projects that deliver on their promises is a positively deviant act! I'm a big believer in positive deviance. Are you?'

Blake smiled. 'Since you put it that way—yes, I am!'

'But do you remember what followed my assertion that 85% of projects fail?' Marguerite asked.

'Of course. *85% of projects fail to meet their sponsors' expectations.* That alleviated my concerns slightly.'

'Because?'

'It emphasized the primacy of the project sponsor. Until I came across that idea, I thought I was the sole arbiter of the project's success or failure. This statistic put me in my place.'

'Excellent. Maxim Two: *There is only one reason for a project: to create something both essential and exceptional.* Why does this matter?'

'Well, to keep the project focused, I guess.'

'How does this maxim achieve such a focus?' Marguerite asked.

'Well, it reminds us that a project has to be transformational—'

Marguerite cut in. 'Does it?' she asked.

Blake frowned. 'A project has to be both *essential* and *exceptional*. To me, the sum of those parts is *transformational*.'

'Hmm. I hear where you're coming from,' Marguerite admitted. 'But before you land on *transformational*, consider *essential* and *exceptional* individually. Is your project *essential*?'

'If you consider saving lives and lowering healthcare costs to be essential, then yes.'

'Good. And is it *exceptional*?'

'We're attempting something that has never been done before. We're constantly drawing from cutting edge research in genetics, disease prevention and computer technology. So yes, I believe it's exceptional.'

'I agree,' Marguerite said. 'I noticed you enjoyed your walk earlier this evening?'

'True true. It left me feeling energized and refreshed.'

'So let's say you asked Taybridge to build you your own private Zen garden near your office. Would that qualify as a project?'

'What is it with you?' Blake wondered out loud. 'How come you can read my mind so easily?'

Marguerite smiled. 'Don't take up professional poker, Blake,' she advised. 'So: Zen garden. Project or not?'

'Well, by your definition, no.'

'Stay with my definition, Blake. It emphasizes the risks involved. Say you bring in some landscapers to create a Zen garden for you. They have to design the garden, which demands considerable creativity, and a deep knowledge of planting and water flows. Then they have to build it, which involves a whole bunch of logistics. Along the way, there'll be some problems for them to solve. For your landscapers, yes, it's a project. For you, it's just a task you delegated to some talented people. It's not a project from my perspective, because the end result is never in doubt.'

'So risk is essential to any project?' Blake asked.

Marguerite nodded. 'A project aims to achieve something that business as usual cannot. But it's risky. Which creates another problem: the human need to minimize or eliminate risk.'

'And yet, you said 85% of projects fail,' Blake observed. 'So you can never eliminate all the risks.'

'Indeed. Some people are so intent on eliminating risk that they end up eliminating any chance of success. They focus on processes, believing that if they follow them religiously, they'll meet with success. I use the word *religiously* for a specific reason. Faith in a project management methodology is as flawed as a primitive belief in a fertility god or goddess. Maxim Three: ***Processes neither make nor break a project. People do.***'

'At the same time,' Blake said, 'you do need some kind of structure.'

'As long as the structure supports the people,' Marguerite said. 'If people have to bend themselves to fit with the method, then the project is unlikely to succeed. There's a simple reason for this. If your best people are focused on ticking a whole sequence of boxes, they're not bringing their individual or their collective genius to bear.'

'That makes sense to me,' Blake said. 'After all, you've used a number of processes over the last two days. Without them, we'd have talked around in circles until we dug our way through to the Earth's core!'

'You're right, of course,' Marguerite agreed. 'But consider Maxim Four: ***A vivid, shared vision unites a team***. That's why I focus on building a collective vision, on understanding the needs of the individual and the team. But before we can do that, there's an essential prerequisite.'

'Talking with Meadows,' Blake mused. 'Building a shared vision with him which we can then take to the team.'

'And why did I insist you talk with Meadows?' Marguerite asked.

'Well, it's important to have him on board, I guess.'

'Why is it important?'

'I'm not sure. Professional courtesy?'

'It's much more than professional courtesy,' Marguerite said. 'Aligning your vision with Meadows' reflects the reality of Maxim Five: *All projects are political.*'

'Political? I see myself as an apolitical being!'

'As do many of your peers,' Marguerite agreed. 'So you and Meadows are besties forever?'

Blake decided a cautious response was required. 'Well, he's not of my generation. But we do get on fairly well.'

'Do you consider him your equal?'

'My equal? Well, he's not much of a coder...'

'So what does Meadows have to offer you?'

'Over $100 million from his group of partners. All up, $200 million, if you count the other VCs he's brought into the fold.'

'Does that money matter to you?'

'I couldn't build Med•evolv without it.'

'Could Meadows find another start-up to take his money?'

Blake considered this unpalatable question for a moment. 'Undoubtedly,' he conceded.

'Now, you and I both know that you are integral to the success of Med•evolv. But the simple truth is you need Meadows more than he needs you. In business, money is power. So I asked you to speak with Meadows because he's your project sponsor. He decides how success looks, sounds and feels. He's the one who decides whether you've succeeded—or if you have failed.'

Blake pondered Marguerite's comment for a few moments. 'So when you say all projects are political, you really mean that to Med•evolv, Meadows is like a king? That I need to defer to him?'

'That's the reality of the situation.'

'But he doesn't expect me to bow and scrape!'

'No. But he expects you to provide a return on his investment. While he believes you're on course, he'll support you. But there are some subtleties you would be wise to observe, which we'll cover later.

'First, though, let's review the steps I advised you to take that day we sat in my lounge room and scattered sheets of chart paper all over the floor. There's value in you knowing this, because your career is going to hang on your ability to manage your sponsors, and on your ability to achieve something essential and exceptional through your projects.'

'You make my entire life sound like a highwire act,' Blake complained.

'Isn't that what you chose?' Marguerite asked. 'These maxims will help you maintain your balance. What was the first thing I told you about the project sponsor?'

'Talk to the organ grinder, not the monkey.'

'An old truism,' Marguerite replied, 'respectful to neither organ grinders nor their impish assistants. But useful nonetheless. Wherever there's money you'll find noisy people clamoring for attention. Some of these might hold out the promise of easy capital when they're merely commission-only agents. Avoid them; speak with the organ grinder.

'We both know the depth of Meadows' connections—with merchant bankers, superannuation funds, and foreign capital. But had we not known, we would have validated his status before approaching him. So having decided to speak with Meadows, what did you ask him?'

'I asked him to describe the end of the project,' Blake said. 'I asked him to *describe success. How would it look? How would it sound? How would it feel?* I asked him to *be as vivid and as concrete as possible*.'

'How did he react?'

'At first, he seemed bemused. So I followed your advice. I told him I wanted to build a rich, tactile image of success. Something we could both work toward. Something clear and measurable. Something I could share with the project team, to bring them on board.'

'And that convinced him?'

'Absolutely. He described it fluently and powerfully. He referred to the scenario of Gina. "When we launch," he said, "we will have one million Ginas signed up and ready to go. The product design will be impeccable: something Raymond Loewy or Jonathan Ive would be proud to call their own. The technology will keep Gina in the loop at all times. And our market capitalization will have trebled."'

'Did any of this surprise you?' Marguerite asked.

'The Raymond Loewy comment, for sure. But Riley put me in touch with a young team of designers taking the West Coast by storm. They've already provided some preliminary sketches that make the Coca-Cola bottle look frumpy. And his throwaway line about keeping Gina in the loop at all times gave us an opportunity to link the system to smartphones and smartwatches.'

'You explored that with him first?'

'I did. And it was obvious to him that knowing when the sequencing will be completed is essential to him. So now it's in scope.'

'Before you closed the conversation, you did something else,' Marguerite prompted.

'Yes. I invited Meadows to open our retreat. If possible, I added, he'd be welcome to return for the close of the first day. I couldn't believe it when he said he'd like to stay for the second day, provided he wasn't imposing.'

'Which tells me he has a deep commitment to this project. Let's recap the steps we took over these last two days.

'First, Meadows spoke about his vision. Afterwards, he took questions from the floor, some of them challenging. He answered them straight, not like a politician.

'Second, I asked people *if they could see, hear and feel the same outcomes* as Meadows. They needed time to soak it in, and they needed the opportunity to add their own embellishments and thoughts—provided they remained consistent with Meadows' vision. So I asked them *if there was anything they would like to add or clarify*. If there were any ideas which ran counter to Meadows, or questions we could not answer, I *parked them on the sideboard*, and promised to return to them later.

'Then I asked, *What do you need to make this vision come to life?* This sparked a storm of ideas that carried us through that dangerous period after lunch, when half the room is ready to fall asleep. These ideas coalesced into *three distinct plans*. I must admit I encouraged this process. Why did I want three plans?'

'As I recall,' Blake said, 'you wanted to ensure buy-in. Developing three different plans meant we explored as many options as possible, and gave everyone a chance to be heard.'

'Some managers believe this to be a needless distraction,' Marguerite said. 'In my experience, however, I've found that the time spent up-front encouraging buy-in to be minimal compared with the time lost when project team members disengage later. As soon as the going gets tough, people tell themselves *it's not my project*, and lose commitment. Treating them as important partners in the planning from the start forestalls this emotional absenteeism later on.'

'At the end of the day,' Blake continued, 'we presented our plans to Meadows. There was something magical about the processes you used, because *we coalesced three plans into one*. And as we did so, a clear consensus emerged. How did you do that?'

'For one thing, I was fortunate,' Marguerite admitted. 'I was working with a group of adults. Sometimes, you have a group that seems adult, but turns out to be adolescent, or infantile. When people are committed to a project they treat each other with greater respect.'

'It wasn't just luck,' Blake said. 'You worked with the group to build a shared vision.'

'I did. And that certainly helped. As did having a roomful of mature professionals, rather than a roomful of schemers and misanthropes. I cannot thank Thomas enough for speaking up at the start. Had he held his tongue, there would have been an undercurrent of discontent within the team that could have derailed the whole process.'

'I thought you were extremely brave,' Blake volunteered. 'You handled dissent elegantly.'

'A skill born of long practice, and motivated by a fear of failure,' Marguerite said. 'Such prejudice needs to be confronted strongly. But I don't want to dwell on that. There was something else I did throughout the retreat that aided the group.'

'I was watching closely, but I must have missed it,' Blake confessed.

'I encouraged people to listen, and to show they understood before continuing. Listening builds shared understanding, and paraphrasing builds rapport. Often, we're so busy planning what we'll say next that we have stopped listening. I worked with the different teams to minimize the chance of that happening.'

'Well, it certainly worked for me!' Blake admitted. 'There were some people there whom I heard clearly for the first time in my life.'

'That's useful feedback for me, Blake,' Marguerite said. On the second day we developed the Green Plan. As you might recall, I broke the room into work groups, and gave them four questions to consider.

'One: identify each task that needs to be done.
'Two: how long will each task take?
'Three: what resources do you require to support these tasks?
'Four: are there any prerequisites and interdependencies you need to consider?'

'That second question confused me,' Blake said. 'Why ask how long a task will take, rather than asking for a specific end date?'

'Because you can only identify the end dates once you have mapped out the project's critical path. If you fix the dates too early, people focus on the dates, rather than on the conversations they need to have to plan the project fully.'

Blake tapped a finger against his chin. 'You know, I'd never considered that, but you're right. I hadn't realized how much psychology is involved in running a project!'

'You'll realize so much more by the time you're done,' Marguerite promised. 'Before we move on to defining our critical path, there are three more points to review.

'One is the use of the **emissaries**. There is no point in a team working in isolation. ***These emissaries need to continue their interdisciplinary work back in the office***.

'The second was my **Quilting Session**. If everything has been working well, then all those different roles and needs will weave together easily at the end of the retreat. If you encounter obstacles, that will tell you that something has been missed, that issues arose earlier in the retreat that were not adequately addressed. Or worse: there are destructive issues stalking the team that remain hidden below the surface. Issues that never surfaced during the retreat. I'm pretty sure we knocked those on the head from the start.

'The third and final point was my use of **question marks**. These are not markers of difference or argument. They just indicate an area where a little more clarity or information is required. Usually, I can tell the difference between a missing piece of data and a buried conflict. Question marks are nothing to fear.'

'I'm glad to hear it,' Blake said. 'And I have a pleasant surprise for you right now.'

'Which is?'

'We can save some time with the critical path next week. Taybridge gave me this software that will do that for you. We just need to punch in some of the key milestones.'

Marguerite shook her head sadly. 'The critical path is too important for that. It maps all the steps we need to complete to finish the project on time. The software I use for generating the critical path is between my ears—and yours. Any time I've seen a computer generated critical path, it's left out some critical issues. That alone should be enough to exclude it, but there's another reason why we need to map this out by hand.'

'What's that?' Blake asked.

'As the project leader, you need to know every nuance of the project. Once you leave it up to the software, you lose touch with the intricate detail. By all means record your critical path on the computer. But always use IT as an assistant, rather than a crutch.'

$\mathcal{N}$INE

Marguerite stood in front of the smartboard, her marker poised like a knife. The team had reconvened at the El Paso Motel with most of the question marks ironed out of their plan. 'You now have a name for your personal genome sequencer,' she began. 'The inGenie. Now it's time to map the project's critical path. Can anyone tell me why mapping the critical path is so important?'

An awkward silence followed. Marguerite allowed the silence to deepen until she had gained everyone's attention. 'Let's say you're running a circus,' she began, her voice deadpan. 'You've hired a train to transport your act from one side of the country to another. The journey should take three days. How long should you allow for your critical path?'

One of the programmers raised her hand. 'Three days?' she suggested.

'So you might think,' Marguerite continued. 'But two hundred miles into your journey, you're planning to meet with another train carrying the lions, which is coming from a branch line. But maybe the lion tamers had the flu.

They've fallen two days behind schedule. So your circus train has to wait at the junction for two extra days.

'Three hundred miles further on, there's another station where you're taking the acrobats on board. Now the acrobats have to cool their heels, waiting on the lions. As will the elephants four hundred miles down the track, then the aerialists, the Wild West show, and the clowns. Every one of these acts is scheduled to meet the train at different times along your route. Any of them could delay you further, and all of them could be inconvenienced by any delays.

'By the way, as well as the cost of hiring the train, you're paying everyone's wages from the moment they reach their pick-up point at the scheduled time. That's why we need to map out the critical path. Allowing for all contingencies, it's the least amount of time needed to reach your end goal. So: let's get started!'

On the right-hand side of the smartboard, Marguerite wrote up 'inGenie launch'. Underneath, she added the following dot points:

- bubbles
- universal acclaim
- 300,000 lives saved in first three years
- exhilaration
- exhaustion.

'These were the key descriptors of success you identified as a team,' she said. 'All we need to do now is work out how we move from here'—she gestured to the left-hand side of the board, which remained blank—'to here.' She tapped her pen on the right side of the board.

'To achieve this, we'll work backward from our end goal. What's going to happen immediately before you launch?'

Wilson called out from the back of the room. 'Someone has to ice the Krug!'

'Very witty,' Marguerite replied. 'You want this on the critical path?'

Wilson tilted his head sideways. 'Not front and center,' he admitted. 'But it needs to be done.'

Marguerite glanced across at Taybridge. 'Aiden?' she asked.

'We've already booked a venue,' he said. 'Which means our launch date is locked in.' A murmur of excitement rippled across the room. 'Don't worry about the launch logistics. We have all that under control.'

'Thanks, Aiden.' Marguerite turned back to the group. 'So what's the penultimate step in the project?'

Blake raised his hand. 'Ramping up manufacture to have half a million pieces ready to shift within one week of the launch date,' he said.

'Great!' Marguerite captured the thought on the board. 'And before that?'

'Final user testing and debugging before manufacturing goes green,' one of the engineers suggested.

'Two excellent points,' Marguerite said. 'Let's just tease them out a little more...'

Over the course of the morning, Marguerite created a chart that began two years in the future, and finished in the present. Or, after the team had reviewed it, a plan that began in the present, driving them inexorably to a fixed moment two years hence. For the first time for many, the immensity of the task they had set for themselves began to sink in.

Again, Blake watched with awe as Marguerite wove together all the information the team provided. But this time, instead of encouraging wild ideas and speculation, she drew everything together, linking all the critical components of the project. She nailed down the timeframes for each chunk of the project, and placed a name against each one—the person who was responsible for either the success or failure of that task. Blake winced each time his name appeared. There could be no denying his accountabilities.

Marguerite placed her marker on the tray beside the board. 'OK, team—how many of you have watched those reality cooking shows on television?'

A few brave souls gingerly raised their hands.

'You know when the host tells the contestants: *Your time starts now*? Well, your time starts now. It's a two-year sprint to the Krug!'

As the team filed out to board the bus, Marguerite gestured at Taybridge to wait. He sat down beside Blake on the sofa, a look of apprehension on his face.

'You two have a problem,' she began, speaking softly. 'I know how much work you've done to bring the project to this point. I know how you've challenged each other, and I know you've been forced to reflect deeply on your preferred style of leadership. I also know you've grown considerably over the last few weeks. Which fills me with hope for the future.'

'Sounds like you would like us to change even more,' Taybridge said, his voice wary.

'It's not what I would like,' Marguerite said. 'It's what the project needs. Aiden, are you OK if I start with you?'

Taybridge swallowed hard. 'If you must.'

'It's daunting,' Marguerite acknowledged, 'but I know you can take it. I've watched you moving between the sub-teams over the three days we've spent together. Blake has briefed me on the work you've done to set up and resource the project. I've heard about the IT systems you've installed, the people you've hired, and the documentation you've commissioned to aid the transition from prototype to manufacture. There's just one question I need to ask. *Are you doing all you can?*'

Taybridge stared at Marguerite. 'No,' he said at last. 'I have a lot more to offer.'

'Go on,' Marguerite encouraged him.

'I feel I've been playing catch-up to whole time. The sign on my office door says *Project Manager*. But I've not been leading the team. I've been following them.'

'Astute observation,' Marguerite said. 'Over the last thirty years or so, the profession of project management has been working overtime to redefine the role of project manager. When you're compiling those endless reports to satisfy the management board, you know what you're doing?'

'Sitting in my office staring into a computer monitor, rather than leading the team?' Taybridge guessed.

'Correct. But it's even worse than that. You've been looking in the rear-view mirror, rather than looking forward. All those reports deal with the past. As project manager, you need to focus on the end goal—and helping the team reach it.'

'Wow!' Taybridge exclaimed. 'So that's why I never seem to gain any momentum.' He paused, and scratched the side of his head. 'But how can I change that?'

'Simple,' Marguerite said. 'No more Percentage Complete Reports, no more Red Amber Green Reports.' Sensing Taybridge was about to interrupt, she raised her hand. 'That doesn't mean you won't be able to measure your progress, doesn't mean you're no longer accountable for results. It simply means you'll have useful information at your fingertips. Information that keeps the project on track.'

'But Percentage Complete and Red Amber Green are standard reports!' Taybridge objected. 'They're what the Board expects!'

'Indeed,' Marguerite nodded. 'And at the same time, you're an intelligent man. Can you tell me why Percentage Complete and Red Amber Green don't serve either you or the Board?'

'In essence, you're asking me to critique my profession,' Taybridge said.

'No profession can survive without self-reflection,' Marguerite replied.

Taybridge swallowed hard. 'Fair enough. Here's something that has been troubling me for a while. Percentage Complete assumes every percentage point carries the same value. But you might have ninety-eight low risk tasks, and two critical, high risk tasks. If those two fail it doesn't mean your project is 98% successful. It means the project's a complete bust. At best, Percentage Complete Reports are a distraction; at worst, they can lead you over the cliff.

'As for Red Amber Green, I'm not so sure. I take your point about looking backward. But if they alert the Board to potential problems, surely they're useful?'

Marguerite nodded. That did not indicate her agreement, only that she'd heard. 'When we're habituated to use certain tools, they seem natural,' she said. 'My problem with Red Amber Green is threefold.

'First, they infantilize the work of project management. No other profession I'm aware of tries to distract its sponsors with colorful lights and bouncing balls. Can you imagine a lawyer presenting a judge with a report that looked like a page from a picture book?

'Second, most of the data are irrelevant. Green means go. So all that green creates the impression that everything is OK. But you know that any astute Board members are going to home in on the red and amber. So why not just get to the point? Answer: because a field of green enables the project managers to feel better about themselves. What's a splash or two of red in a multipage report? But how the project manager feels is irrelevant. It's more important to identify and fix problems *before* they occur.

'Which leads me to my third point. If you're walking down the street and someone falls over in front of you, how do you respond? Do you help them up, or do you write a report? You pick them up, of course. If you can see that some damaged pavement caused the fall, you might report the problem to city hall. But then you're looking forward: you're doing your best to prevent another person from falling in the same spot. Even better if you can fix the pavement *before* someone trips.'

'You feel strongly about this,' Taybridge observed.

'You're a professional,' Marguerite said. 'No one's going to value you unless you value yourself. Forget the paint-by-numbers and the rear-view mirror. Show the Board you're driving the project forward by anticipating problems, and addressing them before they reach crisis point.' Marguerite glanced across at Blake. 'Can you support Aiden in this?'

'Absolutely.' Blake turned to Taybridge. 'What else do you need from me?'

Taybridge took a deep breath. 'Well, I've noticed something that could cause problems for the project. You feel most comfortable with the coders—I guess because it's your area of expertise. But other people, equally qualified and skilled, may feel excluded. I believe you could be more inclusive as a leader.'

Blake pondered Taybridge's request. He glanced across at Marguerite. 'This squares with your observations, doesn't it?'

Marguerite nodded. 'It does. You know it's time to step out of your comfort zone and act like the project leader, not a coder. Which means accepting several key accountabilities.

'Your most important role is to keep the team focused on the end goal, to inspire them to stretch themselves as far as possible. Everyone is working toward an ideal that you first conceived. So you are a beacon to them.

'You're also the one people will turn to when they encounter problems. It's not your job to solve their problems for them. Rather, show them how to solve the problems themselves. Show them how to collaborate with each other. Once you put your pride to one side, you'll realize that often their solutions are better even than yours.

'You also enjoy a close relationship with Meadows. So if you encounter real problems, you're the man who has to manage Meadows. Say the money starts to run out for some unforeseen reason, or one of your expected technological breakthroughs does not materialize. You'll need to initiate an extremely difficult conversation with your primary sponsor.

'Can I give you a key piece of advice here—Aiden, this is important for you, too. If you ever have to approach Meadows, or someone else who wields great power over your project, remember *all projects are political*. You're dealing with an alpha personality, male most likely. Now, while everything is going swimmingly, it's easy to assume he's your best buddy. However, that can all change in an instant. The moment you go to Meadows and say *I need* or *I want* or *I feel*, you run the risk of shutting him down. He doesn't care what you need, or want, or feel!

'Blake, I can see you're shaking your head. Trust me on this. Alpha males have no interest in mothering you. Maybe you've gone cap in hand to Meadows in the past and survived. But if you go to him and say *I need more money*, be prepared for an explosive reaction.'

'So if I do need more money, how should I approach Meadows?' Blake said.

'Try this script instead. *We've reached a critical part of the project. A, B and C are all ahead of schedule. There's an issue with D, and we have several options. The team would value your input.*'

'And the options are?' Blake asked.

'*Settle for a lesser outcome. Extend the project timeframe. Buy in the IP from outside the company.*'

'But I'd have already decided on my preferred option?' Blake said.

'Naturally. And if Meadows disagreed, you would make your case objectively. You would then accept his final decision.'

'Even if it compromised our final outcome?'

'In that case, you would realign your vision with Meadows', and inspire your team to achieve that revised goal.'

'Even if they were unhappy with the revision?'

'You're the project leader, Blake. It's your job to address any unhappiness, and turn it into something far more positive.' Blake looked ashen. 'That's when you are most tested as a project leader. And when your relationship with your project manager comes under the greatest scrutiny.'

Taybridge frowned. 'Because all our team members will be watching to see whether or not we're united?' he asked.

'Precisely!' Marguerite offered Taybridge a luminous smile. 'It's essential the two of you show the rest of the team how productive professionals cooperate with each other. Every single thing you do, either as individuals or together, must be directed toward achieving that final vision. It means making sure your team is on task. And, more importantly, making sure every relationship within your team is 100% functional.'

'So there's no room for the conflict you often see in teams?' Blake asked.

'Conflict?' Marguerite said. 'Depends how you define it. No team can bring a product as sophisticated as inGenie to market unless all team members are willing and able to engage in robust professional conversations. By which I mean: if something you've done is suboptimal, be prepared to listen to feedback from your colleague. If you advance an idea within your team, expect it to be stress-tested. And if a co-worker suggests minor but useful improvements, engage with them and bring those improvements to life.

'At the same time, there is no room on a team like yours for backstabbing, petty jealousies, or *ad hominem* attacks. There's no room for people who can't take constructive feedback, or who bring their personal psychological dysfunction to work. Managing people and their interrelationships is the key priority for both of you. It's a delicate balancing act: while all projects are political, you can't let negative politics permeate your team. You manage the politics, so your people can build something brilliant.'

Taybridge grimaced. 'I only wish someone had taught me this ten years earlier,' he said.

'Let me paraphrase Dylan,' Marguerite suggested. 'You're a project management artist. You don't look back.'

Taybridge shook his head. 'An artist?' he muttered. 'I don't know about that!'

'A project manager is an artist,' Marguerite insisted. 'There's no alternative. Sure, there's a science to it, but there's also an art. You have to be both: artist and scientist. Use your intuition and personal flair; use logic and measurement.' She paused, and softened her gaze. 'You're feeling uncomfortable right now,' she observed.

'Well, in a couple of weeks I've gone from compiling Percentage Complete and RAG reports, to something much more ambiguous, much less certain.'

'And that's the biggest risk this project faces,' Marguerite said, her voice low and soft. 'You returning to your comfort zone, rather than confronting this uncertainty head-on.'

Taybridge slumped back in his chair and rubbed his hand across his mouth. 'Wow,' he said. 'That really ups the ante!'

Marguerite turned to Blake. 'There's something you can do for Aiden that will push his artistry to the limits. Given the complexity of your project, it's essential. It will help you manage your collective risk. It will also require every last ounce of Aiden's creativity and courage.'

'Sounds confronting,' Blake remarked. 'What is it?'

'Simple,' Marguerite replied. 'Break your project deliverables down into the smallest possible steps, and *test every step*. Someone writes a discrete chunk of code. Test it. Test it in conjunction with related pieces of code. Test it in conjunction with the hardware. And always follow this essential rule: *test with real data*. This point is essential. Real data contain the errors and glitches that will push your system to its limits. If you use data that have been homogenized and pasteurized, your testing will only reveal how the system functions under ideal conditions. And real life, as we know, is often far from ideal.

'So: an example. Say you have written some code that reads data from a chemical sensor. Does that code run perfectly itself? Does it work in conjunction with the sensor? And does it work with the code that compiles that data to send to your database?'

'That certainly doesn't fit with the phases and tollgates used in Prince2,' Taybridge volunteered. 'However, it sounds sensible to me—sensible and time-consuming.'

'It will take more time,' Marguerite admitted, 'until you reach the end of the project. Back when I was working toward my PhD, I interviewed a crusty old IT project manager who worked for a nameless government department. On his wall he had a framed photocopy—actually, an analog copy of an analog copy several generations old—which summed up his entire professional experience. It showed two ants standing in front of a complex project chart—a real spaghetti tangle of boxes and diamonds. On the right side of the chart was a box marked *End*. The penultimate box was marked *Then a miracle occurs*. One ant was pointing to this box. He said to his colleague, "I think we need a little more detail right here."'

Taybridge laughed out loud. 'That's funny because it's cruel,' he said. 'But I thought ants were so much capable than us.'

'They do seem to manage their projects more efficiently than humans,' Marguerite admitted. 'Which is probably why the cartoonist decided to depict ants, rather than people. It makes the truth much less threatening.'

'So by testing each step, we reduce our reliance on a last gasp miracle?' Blake asked.

'Exactly,' Marguerite said. 'If you test each step separately, it's easier to pinpoint failures. But you also need to conduct rolling cumulative tests. Keep adding in. If you wait until the last minute to bring everything together, your project could prove impossible to debug. And that's when you'll find yourself praying for a miracle.'

'So while we invest more time during the project, we make that up in the closing stages,' Taybridge said.

'That's been my experience,' Marguerite replied. 'And as project manager, it's your job to track each test, check it off as successful, and prepare the team for the next test.'

'Or mark it as a failure, and send it back for revision,' Taybridge added.

'Absolutely,' Marguerite agreed. 'Without shaming anyone for the failure, or creating a sense of panic because your timeline may have slipped slightly.'

'Sounds like there's so much to balance,' Blake said. 'What if we—what if I—screw something up?'

'Acknowledge the mistake,' Marguerite said. 'Learn from it. Move on. If your team sees you self-flagellating because of a relatively minor mistake, they'll lose faith in you.'

'But self-flagellation has been one of my core skills!' Blake objected.

'Tell me, Blake. Did Meadows give you the name of a therapist?'

Blake felt a need to correct Marguerite. 'Actually, she's a high performance coach,' he said.

'So Meadows gave you her name?'

'Yes.'

'Have you seen her?'

Blake glanced across at Taybridge. He felt very exposed. 'Not yet,' he confessed.

'Do it today,' Marguerite commanded.

Blake stared down at his feet. He did not appreciate Marguerite's tone of voice. But before he could object, Taybridge spoke up. 'A therapist slash business coach? Can you give me her contact details, Blake?'

'You don't feel ashamed to ask?' Blake said.

'Not at all,' Taybridge replied. 'I'd feel ashamed if I didn't seize this opportunity.'

Marguerite offered Blake the faintest of smiles. 'This project is yours to run,' she said.

'Meaning?'

'Make sure you have all the resources you need.'

'I'm not sure I need a therapist,' Blake said. 'And as for a high performance coach...'

'Don't you see yourself as a high performer?' Marguerite asked.

Blake glanced over at Taybridge. 'Maybe we'll be competing for appointment times,' he said.

Marguerite began to pack up her markers. 'I'll scan all the notes from the smartboard, and my notes from this evening. You'll have them first thing tomorrow morning.'

'So now that we've done this work, the rest of the project should run smoothly?' Blake asked, a note of optimism in his voice.

'Not quite,' Marguerite replied. 'Now that you've done this work, you've given your project an even money chance of success.'

$\mathcal{T}$EN

'Long time no see!' Riley embraced Blake on the pavement outside The Creamery. 'I hear you're in the business of anticipating miracles now!'

'Forestalling the need for them, more likely,' Blake said. 'And how did your disagreement with Littler play out, Ms. Pearce?'

'I did what Marguerite suggested, and now he's eating out of my hand. He's already planning to launch six new banking enterprises across the Pacific Rim. And he's offered some great red-tape busting ideas to our project. Turns out he wanted to be a part of this more than anyone else.'

'Great. So you're on track?'

'I was, but I had a bit of a hiccup earlier this week.'

'Before you tell me the story, can I order you a double-shot ristretto latte?'

'If you do, I'll love you even more.'

'Take a seat while I take care of this,' Blake suggested. He watched Riley push through the crowded café to their favorite spot in the corner.

'So—just a hiccup, or something more?' Blake asked.

'Just a hiccup,' Riley said, 'although it could easily have turned into something catastrophic. Graeme, my best business analyst, has a bad habit—he loves skateboarding. On the weekend he hit a curb, fell, and smashed not one but both of his wrists. Twenty-four hours after the ambulance took him to hospital he checked out with plates and pins in each wrist, and two months off work.'

'Ouch, ouch and ouch,' Blake said. 'Ouch for the fall, ouch for the surgery, and ouch for the trouble this creates for your team.'

'Turns out there were only really two ouches,' Riley said. 'Ouch when Graeme fell, and ouch for his health insurance company when it received the bill. The effect on our team has been minimal.'

'But he's your top business analyst,' Blake said. 'Or don't business analysts matter?'

'I thought Graeme was pretty much indispensable,' Riley replied, 'and I'm keen for him to return. He's the interface between our project and the different business arms of the bank. He works out their business needs, and communicates them to us. Without him, our lawyers would be unable to develop new internal protocols.'

'Sounds like you really rely on him. So how did you cope when he called in?'

'His partner called, actually. At the moment, Graeme can't even pick up a phone. But once I heard what had happened, I followed Marguerite's advice.'

'Which is?'

'First up, *remain calm*. Pretty standard, but you'd be amazed how many people go into crisis mode the moment they receive bad news. As Marguerite says, *project managers instill a sense of calm*. Next, *inform the team*. Traditionally, managers try to solve such problems, then tell the team once they've slotted someone else into Graeme's vacant spot. Not Marguerite. She informs the team before addressing the problem.'

'Why's that?' Blake asked.

'Because Marguerite knows the project manager won't have all the answers. The solution you or Taybridge come up with may not be the best one for the team. Why? Because the team members understand the intricate details of their work better than you ever can. They also have a much finer appreciation of each other's strengths and weaknesses than you do. So here's Marguerite's iron rule of problem-solving: *involve your team*.'

'That's all very collegiate,' Blake replied, 'but what if the team can't decide?'

'Marguerite does not say *let the team decide*. She says *involve the team in solving the problem*. There's a big difference between the two.

'Here's how I handled this specific problem. I told the team Graeme would be out of action for two months, and that we had to work out how to cover his absence. I then asked for ideas from the team.

'It took maybe twenty minutes or so, but a clear consensus emerged. Move George, who had been Graeme's understudy, into Graeme's position for the next eight weeks. While he's not as experienced as Graeme, George has solid analytical skills. Divide George's other work among the four other members of the team. It became obvious we'd built some resilience into the team, because the four were easily able to take on the new responsibilities. After the redistribution, there was still some work left over, so we agreed to hire a temp to cover it.'

'A fairly well-skilled temp, I assume?'

'Sure,' Riley agreed. 'We need a strong skillset. But the rest of the team has covered the more complicated tasks. I've just met with our agency to narrow the shortlist down to three.'

'So how long did it take for you to get on top of this?' Blake asked. 'The guy only stacked his skateboard a few days ago.'

'You need to move fast when these problems arise,' Riley said. 'If I left it a week, I'd not just be down five days of Graeme's contribution. His absence would leave the rest of the team underutilized. As it was, we'll still be playing catch-up for a week or so. However, by involving the team, I've minimized the amount of time that would be lost with people gossiping in the tearoom, complaining about a lack of direction.'

Their coffees arrived. Deep in thought, Blake sipped at his mocha. 'What if you and the team had disagreed on the best solution?' he wondered out loud.

'As the project manager, it's ultimately my responsibility,' Riley explained, 'and, therefore, it's my call. I always listen to my team. If I disagree with their views, I explain why I've made a different decision. So far, I've not encountered any real pushback.' She glanced across at Blake, who could not conceal his disagreement. 'I know it runs counter to conventional managerial wisdom. But if you treat your team members as pieces to be moved around a chessboard, don't be surprised if they lose faith in you.'

Blake could not resist the urge to argue. 'Back in the early days of Pyrouette,' he said, 'I had to roster staff for overtime. If I gave everyone exactly what they wanted, I'd never have finished the list. I had to present that roster as a *fait accompli*: either like it, or leave!'

'Sure,' Riley said, 'some things can't be easily accomplished by consensus. But remember that Marguerite's ideas are part of a complete project management ecosystem.

'Everyone on the team is focused on achieving a shared goal, one in which they believe passionately.

'Each team member knows his or her role, knows what they need to achieve, and by when.

'Each team member knows the rest of the team is depending on them.

'The team has been selected to ensure its resilience. You've chosen people who can work together professionally, people with complementary skillsets. You've also chosen people who are adaptable—in a pinch, they can switch from one job to another at short notice.'

'Marguerite never mentioned anything about building resilient teams,' Blake objected.

'Correct. And you know why?'

Blake shrugged his shoulders. 'No idea,' he admitted.

'Because Taybridge, either by accident or by design, had already selected a team that largely met Marguerite's criteria.'

'I doubt it was by accident,' Blake said. 'Choosing the right people takes a lot of insight.'

'It does. And keeping the right people in the right frame of mind requires even more insight.'

'Now you're speaking in riddles again, Riley.'

'Not at all. Selecting a strong team is only half the story. Weak or ineffectual leaders will destroy even the strongest team in less time than it took to recruit them!'

'How so?' Blake asked.

'If the CEO and project manager are unable to channel the team's energies toward a common goal, the team will lose faith. If the CEO and project manager behave unfairly, if they distance themselves from the team, if they don't acknowledge and reward the expertise of their people, the team will begin to disintegrate. Even devoted team members can turn rogue if they see you jetting about in first class and turning a blind eye to any emerging problems.'

'But we haven't done any of that!' Blake exclaimed.

'Correct,' Riley affirmed. 'Which is why Marguerite has so much confidence in Taybridge. Some people would see his personality as rather beige. But he always treats people fairly and respectfully. Even so, he's capable of so much more. Marguerite sees the unfulfilled artist in him. And she knows that if you can liberate his inner artist, you'll smash this project!'

'Sounds like you and Marguerite have been discussing this project in depth,' Blake suggested.

'We have. Her comments on Med•evolv help me perform even better at Bank Pacific West.'

'You're not trying to build a gene sequencer?' Blake asked with feigned alarm.

Riley chuckled. 'That's not what I meant,' she said. 'You can always learn from other people's successes and failures, even in totally different organizations.'

'True true.' Blake stirred his mocha to capture every last drop of chocolate. 'And what does she say about me?'

Riley sighed. 'I knew this was coming! Marguerite says you're the only person at Med•evolv who's truly indispensable.'

A warm smile spread slowly across Blake's face. He felt his pride beginning to swell.

'She also says you're the only person who has the potential to royally screw your project.'

Blake stared at his empty cup. Suddenly, he felt very small indeed.

Riley reached over, and placed her hand on top of Blake's. 'So tell me,' she said. 'How's it going with your high performance coach?'

'Over the weekend, I had a sudden flash of inspiration!' Taybridge stood in front of a wall in the Med•evolv cafeteria, which he'd lined with glossy chart paper. Then he'd mapped out the critical path on those sheets of paper, using simple images and symbols as well as words. 'This is our Bayeux Tapestry!' he announced to Blake.

Blake smiled. 'Just as long as I don't cop an arrow in the eye,' he said. 'Did you check this with Marguerite?'

Taybridge shook his head. 'Why would I do that? Remember the energy in the room last week as we mapped out our critical path? I wanted to keep that energy alive for the next two years, and this seemed a great way to achieve that.'

'You're sure Marguerite will approve?' Blake sounded unsure of himself.

'It's not our job to continually second-guess ourselves, Blake. Remember what Marguerite told me? I'm an artist. I don't look back. Here's how this chart—the inGenie Tapestry—is going to work. As each team member finalizes a task, he or she crosses it off with a red marker. My job is to stay ahead of the red—to make sure everything is in place for the next step, or the next three steps. If any aspect of our plan changes, I adjust the chart. That's why I'm using erasable markers. And by displaying this in our lunchroom I win total buy-in from the team. Already there are people bragging about how they're going to be the first to grab that red marker.'

Blake frowned. 'So all the team members are happy for their progress to be measured in public?'

'Absolutely,' Taybridge said. 'People love to be acknowledged. They love some good-natured competition. We're keeping that start-up hacker spirit alive. To me, it's a win/win/win.'

'And you did this all yourself?' Blake did not mean to sound incredulous. He'd never imagined Taybridge— beige old Taybridge—could conceive of something so outlandish.

Taybridge smiled. 'You're doubting me. That's OK. I doubted myself when this idea first popped into my head late on Saturday evening. Years ago I took an online course in cartooning. Not that I expected to be published in the *New Yorker* or anything; I just liked doodling. I never thought it would lead anywhere, but just look at that!' He gestured toward the wall. 'This is my payoff.'

'I must admit, it is impressive,' Blake conceded. A thought came to his mind. 'You didn't happen to see that therapist on the weekend, did you?'

'Good guess,' Taybridge said. 'I've no need to hide it. Saturday was her only chance to fit me in. Somehow, talking with her fired me up. I guess she just affirmed my faith in myself. Something that no one's going to undo.' He looked directly at Blake. 'No one.'

'I didn't mean to sound critical,' Blake said, by way of an apology. 'I'm surprised, that's all.'

'There's no shame in seeing her, Blake.' Taybridge turned to his chart, and added a little extra shading beneath an arrow. 'Who knows how she might inspire you?'

Blake's head bobbled—a movement halfway between a shake and a nod. 'We'll see,' he muttered. He paused, deep in thought. 'You've certainly made me think. I doubt you're an outlier. Maybe there are other folk on the team with hidden talents.'

Taybridge nodded. 'I'm sure there are. It's my job to discover them. Because the more we can bring their talents to the fore, the stronger the outcomes we'll achieve together.'

$\mathscr{E}$LEVEN

'Unbelievable.   Un-expletive-deleted-believable!'
Blake, Taybridge and Wilson stared at the computer screen. 'Click *Replay*,' Blake demanded.

The YouTube video loaded again. A young girl and her grandfather walked hand-in-hand through a summer meadow, their hair backlit by the sun. In the background, Mozart's *Requiem* soared. Then a voice-over began, in unctuous tones.

'She wants you to be there for her.

'When she stars in her school musical.

'When she competes in her athletics carnival.

'When she graduates from high school, and then from university.

'When she walks down the aisle to marry her forever partner.

'When she experiences the miracle of birth.

'Our lives can be as long as we have dreamed.

'Imagine a world where you can monitor your genetic status at the touch of a button.

'Where you can identify life-threatening illnesses *before* they arise.

'Where you can correct genetic anomalies, quickly and permanently.

'All from the comfort of your own home.

'On February 14 next year, Recombinant Technologies will unveil a revolution in genetic self-care.

'Until then—hold her hand, and tell her how much you love her.'

'*Forever partner*?' Taybridge intoned. 'What is that—some kind of pet?'

'The miracle of birth? Talk about disgusting!' Wilson sneered. 'What kind of girl wants her gramps in the delivery room?'

'I'm not sure that's what was intended,' Blake said. 'But my Lord, talk about an excess sugar hit! That's exactly the kind of advertising I don't want for inGenie.'

Taybridge peered at the screen. 'The video's been up for less than a week. Already it's had 250,000 hits. Trust me—this is going viral.'

'How many members of our team have seen this?' Blake asked Wilson.

'By now, pretty much everyone,' Wilson replied. 'It's gone through the office like wildfire.'

'This could really hurt us,' Blake said. 'Next February is less than ten months away. We're still eighteen months from launch, assuming everything runs to plan.'

'They're not claiming to launch next February,' Taybridge cautioned. 'They're promising to unveil a revolution. If they have a real product in the pipeline, why not show it? Why not set a launch date?'

'You think it's vaporware?' Wilson asked.

'I don't know enough right now to answer that,' Taybridge replied. 'I want to know everything I can about Recombinant Technologies. Who's the CEO, who's backing them, who's working for them. Once we have some real answers, we can assess the threat they pose. Are they offering a revolution, or just fairy floss?'

'Recombinant Technologies,' Blake mused. 'Their name implies they can clone and repair genetic material. If they can, they're well ahead of us.'

Taybridge gave Blake a stern look. 'Remember everything Marguerite taught us. Let's stay calm. Engage with our staff. Focus on our end goal. Until we know who's behind Recombinant, we won't know whether or not they're a real threat.'

'So what do you make of this video?' Blake said. 'What does your gut tell you?'

'I'd say someone is trying to get under our skin,' Taybridge replied.

'You may be right,' Blake said. 'All-staff meeting in the cafeteria. Immediately.'

'You know why I welcome this?' Blake asked his team. 'Because it shows we have something other people want. Here's what we know about Med•evolv. We're focused, we have the best team in the industry, and we're kicking goals. Here's what we know about Recombinant Technologies. Nothing. Nada. Zilch. We'll look into them, find out what game they're playing. As soon as we know, we'll let you know. Until then, none of us has anything to fear.'

As soon as everyone returned to their desks, Taybridge approached Blake. 'I've engaged a corporate investigator. He knows it's urgent. He'll get back to us with the result of his initial search within forty-eight hours.'

'Great. In the meantime, I need to speak with Meadows.'

'This place is legendary.' Meadows glanced around the crowd at The Creamery—young, animated, busy building new worlds. 'I've heard so much about it. But this is the first time I've been invited here.'

'They do a mean mocha,' Blake said.

'That's not my point,' Meadows replied. '*I've heard so much about it.* Which means, before I even walk through the door, I know I can trust this coffee shop.'

Blake looked puzzled. 'How much trust do you need to put into a mocha?' he asked.

'I'm not talking about mochas,' Meadows replied. 'I'm talking about Med•evolv. Broadly speaking, all the players in this city know what you're up to. Not the intricate details, of course. You're managing the confidentiality aspects of your project exceedingly well. But you've engaged genetic scientists and IT specialists. It's not hard to put two and two together.

'On the other hand, I've never heard of Recombinant. Neither have any of my colleagues. That doesn't mean they're fake—perhaps they're able to fly totally under the radar. But we know the leading genetic researchers, and we know which companies they're working for. And they're not working for this lot.'

Blake's smartphone beeped. He glanced at the screen.

'Apparently Recombinant Technologies is a shelf company,' he announced. 'It's registered to an empty office in Pennsylvania. Right next to Valley Forge.'

'Hardly the place to go if you want the brightest minds in IT,' Meadows said. 'Washington camped at Valley Forge during the winter of 1777, his army beset by starvation and disease. They're sending you a message: they might be the underdogs right now, but they believe they'll prevail. Don't worry—I sense a bluff.'

'So if they're not genuine, what's their game?' Blake wondered.

'My initial guess? Someone wants in on your territory. They'll push you and push you to see if you fall over. Now, more than ever, it's essential for you to keep your team together. If someone wants to steal your lunch, they'll need to steal your people first.'

'Take a look at this.' Taybridge ushered Blake into his office, closing the door softly behind him. 'Another video from our friends at Recombinant.'

The vision appeared to have been shot in an apartment overlooking an unfamiliar city. The camera closed in on the face of a man in his mid-thirties. To Blake and Taybridge's surprise, he began speaking in Russian. Subtitles appeared as he spoke.

'The West underestimates Russia at its peril. Our scientists and computer technicians have achieved breakthroughs that our competitors can only envy. Which is why a technology as powerful as Recombinant's could only come from beyond the Black Sea. And to our competitors, we say this: we know all your secrets. What do you know of us?'

As he spoke, he picked up a sheaf of papers, held them up to the camera, and slowly turned the pages.

'Good grief!' Taybridge choked on his cup of tea. 'Those are our scoping papers!'

'Is this line secure?' Marguerite asked.

'I certainly hope so!' Blake spluttered.

'I've just sent you an encrypted message,' she replied. Then the phone went dead.

Blake logged on to Cryptos, and entered his passcode. Without telling anyone, he walked over to the lifts. As he descended, he summoned an Uber.

'Why all the spycraft?' he asked Marguerite. They had taken a back table at a yum cha restaurant in Chinatown. The background noise eliminated any risk of being overheard.

'Here are the possibilities,' Marguerite replied. 'Someone in your office is leaking information to Recombinant. Or Recombinant has hacked your system. Or Recombinant has used some social engineering of its own to steal your secrets. So you need to track down the source of this information. But here's the rub: you need to do it without establishing a totalitarian regime at Med•evolv.'

'You mean coming down hard on people, upping computer security, checking backpacks on the way out?'

'Exactly. How would you describe your security protocols at present?'

'Industry standard. Taybridge commissioned an audit, which found no evidence of hacking, or unauthorized leaks from our system.'

'Nothing has gone out through email or the VPN?'

'Nothing.'

'How many staff had access to e-copies of the scoping paper?'

'Everyone. We encouraged them to download the documents, so they could comment and contribute to their development. The version on the video is Release 0.97, so it's close to the final product.'

'Security on the downloads?'

'Only on password protected laptops. No tablets.'

'USB drives?'

'They have to be password protected, too.'

'Robust passwords?'

'Minimum of twelve letters or numbers, a mixture of upper and lower case. At least two symbols in the mix as well.'

'OK. Here's the drill. Because there's been a leak, tighten up a little. If a document contains information you can't afford to lose, classify it. Classified documents don't leave the office, whether on paper or digitally. But no recriminations for any staff members who accidently slipped up. No witch-hunts. That's what Recombinant want.'

'How so?'

'The scoping papers define your project scope. They don't reveal any trade secrets, correct?'

'Yes.'

'Here's what usually happens. Your enemies force a leak, and reveal it publicly. You overreact, and suddenly your employees feel like they're working under the watchful eye of the KGB. Before long, one or more of them steal your real secrets and sell them to Recombinant.'

'To pay me back for being dictatorial?'

'Exactly. But now's the time to act like Gandhi. Not Stalin.'

Blake was so fully engaged in reviewing code that he did not hear the tentative knock on his office door. Only when the knocking grew louder did he look up from his screen.

'Come in,' he sighed.

The door pushed open. Damien Scott, one of his genetic scientists, poked his head around the corner. 'If you're busy, I can come back later,' he offered.

'No, no. Now is as good a time as any. Please take a seat.'

Damien could not bring himself to look at Blake. Instead, he focused intently upon the floor.

'My carpet's not that interesting,' Blake said. 'Do you have something you want to get off your chest?'

'It's about the scoping paper,' Damien whispered. 'Release 0.97.'

'Yes?'

'A couple of months ago, I met this guy at the Black Bear Lodge. We kind of, you know...'

'Hit it off?'

'Hmm. We ended up back at my place.'

'I don't need to know...'

'I always wear my USB drive on a lanyard around my neck. And each night, I place it in the top drawer of my bedside table before going to sleep.'

'And the next morning,' Blake guessed, 'after your companion left, you couldn't find the drive?'

'I looked everywhere for it. I thought I must have misplaced it, but it never turned up.'

'But you had made a habit of putting it away each evening?'

Damien blushed. 'I may have been a little—ah— distracted by events that night. So I thought maybe I put it down somewhere else.'

'Distracted, you say?'

'I was pretty wasted. By the time I woke up next morning the guy had gone.'

'But the USB drive was password protected?'

'It was.' Once again, Damien took a keen interest in Blake's nondescript carpet. 'But that's the thing about E. Maybe you confide in people when you shouldn't.'

Blake felt a vein begin to throb in the side of his head. 'Stay calm,' he told himself. 'Let it pass.'

'You might have had a hypothetical conversation about passwords with your hook-up?' Blake suggested.

'My memory's vague,' Damien admitted. 'I remember him telling me his password, and how he had constructed it. Not that I believe him now, of course...'

'Was there anything else on that USB drive that could hurt us?' Blake asked.

'You mean like official documents?'

'Anything that could hurt us. Period. But yes, let's start with official documents.'

'Release 0.97 was all I had.'

'Good. Anything else.'

Damien began to shiver. 'There might have been some images.'

'Porn?'

'Yes.'

'Gay porn?'

Damien nodded silently. Blake leaned back in his chair, and breathed deeply to quell the pain rising in his head.

'Photos of you?'

'Uhuh.'

'You know this puts you at risk of blackmail?' he asked.

'I do. But I haven't heard from anyone.'

'Not yet. Can I level with you, Damien?'

'Of course.'

'I'm disappointed with what's happened—but I'm glad you've told me. You're a valuable member of the team, so I'm not going to punish you or yell at you. Just tell me this: who would be hurt if those photos were released?'

'Me, mostly. I'm out, so my parents know about my preferences—not that they'd appreciate seeing those photos.'

'How would you feel if those photos somehow made their way into the public domain?'

'Humiliated. But I'd rather that than lose my job.'

'Just let me know if anyone approaches you. Recombinant is ramping up the pressure, so they could make contact at any time. Now go back and keep doing what you do best.'

Damien stood up to leave. All the gloom had lifted from his face and shoulders. 'Thank you, Mr. Stein.'

'Come on, Damien. My name's Blake. Just one thing.'

'No more E?'

'If it means you can't keep a secret.'

'You needn't worry. I haven't touched it since.'

'So—this is becoming very interesting.' Marguerite rubbed her hands together gleefully. 'And how is Damien to respond if they contact him?'

'He's to let me know immediately.'

'And then?'

Blake scratched the side of his head. 'I'm not quite sure.'

'What are your best options?' Marguerite asked.

'Well, we could feed them disinformation. Something that suggests we're off track.'

'Benefits?'

'Well, it throws them off the scent.'

'Disadvantages?'

'None that I can see. To me, it's all upside.'

'Try these,' Marguerite suggested. 'One: you need to develop the misleading material, and it needs to be sufficiently professional to fool Recombinant. That will require resources you would put to better use advancing your project. Two: you're asking Damien to play the spy. It extends his relationship with his blackmailers. How successful has he been at dealing with them so far? Chances are, they'll find some other leverage over him. Three: you provide them with material they can use later to ridicule you. All it takes is one convincing YouTube video, and Med•evolv becomes a joke.'

'So what are you suggesting?'

'Don't engage. Tell Damien to ignore them. If those photos do come out, he'll just have to deal with it. His mistake, his consequence. Now: when's my mocha coming?'

No sooner had Blake returned to his desk than an email from Meadows bounced into his inbox.

'Heard of the British police drama *Sweeney Redux*? Probably not, because only the pilot was made. Click on the link and go to 11.38.'

Blake clicked the link, and a video opened on his screen. He dragged the slider forward. The camera closed in on a leather-jacketed thug. 'Over by the gasometer,' he said in a thick Russian accent. Then the camera panned across the industrial wastelands, and the policemen's footsteps crunched in the gravel.

'That's our man from the second video!' he told himself. 'So he's just an actor they've hired to play a role. Thank goodness for Meadows, and his love of obscure English television.' He skipped to the credits. *Man in Leather Jacket: Sergey Bulgakov.*

The next day, Damien approached Blake. He'd received a text message on his phone from Recombinant, demanding more up-to-date material. 'We have photos I'm sure you'd rather keep private,' the message read.

Blake took the phone and tapped a reply. 'Happy to swap for our photos of Sergey Bulgakov. Nice green screen work with footage of St Petersburg in the background—shame the lighting betrayed you.'

$\mathcal{T}$WELVE

'Everything seems to be back to normal?' Marguerite asked.

Blake nodded. 'As far as I can tell. Mind you, the whole episode shook me up. Why would anyone pull such a stunt?'

'Good question. We'll discuss that shortly. But in the meantime, let's reflect on how you kept your project on track while under fire. This is important. As things progress, you'll encounter other problems—perhaps from within the team, perhaps from outside. Knowing what works and what doesn't will help you next time.'

'Good point. We'll be less stressed if it happens again.'

'Any future problems are unlikely to exactly mimic this one,' Marguerite warned. 'So I always ask myself this question: what principles can I draw from that experience?'

'As opposed to?' Blake wondered.

'As opposed to developing a specific set of procedures to follow. Procedures that may only work if the same situation is replicated, and maybe not even then.'

'I understand!' Blake experienced a moment of illumination. 'I use certain key principles when coding, such as writing code that is easy to maintain, and minimizing dependencies between different sections of code. Then I have specific recipes I use to achieve specific outcomes—interface with a CNC lathe, for example. You're talking about the former, rather than the latter.'

'Precisely. So, go back to the moment when Wilson first showed you that sentimental video with the grandfather and granddaughter. How did you respond, and how did your response work for you?'

'Well, at first I felt confused, then angry,' Blake said. 'Those feelings only lasted briefly, however. I knew virtually all the members of our team had seen the video, so I called them together to reassure them. I had to let them know I wasn't worried about Recombinant, and I also had to convince them I had the matter under control. I think I succeeded at both.'

'What makes you so sure?' Marguerite asked.

'Neither Taybridge nor I noticed any discernible difference in the mood within our team, or the levels of productivity.'

'Good observations,' Marguerite said. 'And then?'

'I brought Meadows into the loop. It's important I speak with him before he hears about it on the grapevine.'

Marguerite nodded. 'Exactly,' she said. 'It also shows you're confident you have the matter in hand.'

'That's useful feedback, Marguerite. Taybridge then audited the IT system, to see if any classified information had crossed our firewall. He also re-evaluated our security protocols.'

'After which you came and spoke with me,' Marguerite said. 'Knowing you'd already ordered those audits made my job easier.'

'And I appreciated the advice you offered,' Blake replied. 'Especially the suggestion that I tighten security without turning our staff into criminal suspects. I'm sure that getting the balance right there contributed to the positive mood within the team.

'It also helped me when Damien made his confession. In the past, I might have snapped at him. To tell the truth, his stupidity annoyed me—except I know that I've been pretty foolish at times myself. By remaining calm, I won his loyalty. Which meant he told me about the photographs, and the blackmail attempt. If I'd come down hard on him at the start, he may not have admitted to those stolen images.'

'Anything else?' Marguerite prompted.

'Of course!' Blake exclaimed. 'You showed me I would be unwise to play any games with Recombinant. Play it straight, don't give them the upper hand. That was great advice. When they made their next move, I was able to shut them down cleanly.'

'You handled a potential crisis well, Blake. Now tell me this: is there anything you would do differently?'

Blake rubbed his hands over his chin. 'Nothing I can think of,' he said, trying to sound modest.

'Interesting,' Marguerite said. 'When I reflect on my experience, there is always something I can improve.' She allowed the silence to grow between them. Eventually, Blake felt compelled to speak.

'Maybe I could have been more alert to the possibility of industrial espionage from the start,' he volunteered.

Marguerite gave him a look that left him feeling very exposed. 'I guess you could. But just remember: whoever is behind Recombinant is still out there. We don't know what they want. We don't know how or when they'll strike next. But strike they will. Be sure of that.'

'What do you think they're after?' Blake asked, glad to have changed the subject.

'They could be developing a competing product. If so, they're trying to gain first-mover advantage. However, your intelligence suggests this is not the case. So I'd say they're stalking you. Med•evolv is taking some big risks. If you succeed, you'll have a unique product other companies will covet. Perhaps they're trying to shake your tree, to see if any fruit falls into their lap. Perhaps they're engaged in espionage on behalf of a foreign power. Or perhaps they're trying to devalue the input of your VCs, and the perceived value of Med•evolv.'

'But we're eighteen months away from our IPO.'

'Agreed. But anytime anyone acts in ways that seem irrational, follow the money. You'll witness some bizarre behavior at that place where high finance and human greed intersect.'

'Fair enough,' Blake conceded. 'Any of those options sound plausible. But if they want to steal our technology, why wouldn't they wait until we have a final, proven product?'

'Good point,' Marguerite said. 'So perhaps Med•evolv is not their main focus. Maybe they're trying to pressure a member of your board, or maybe even Meadows himself. Remember my maxim about politics?'

'*All projects are political*,' Blake stated.

'So now, more than ever, political awareness is essential to your success,' Marguerite said. 'And remember: the most successful politicians think several steps ahead of their rivals.'

As Blake was walking back to his office after his meeting with Marguerite, his mobile rang. He glanced at the screen and pressed the green button.

'Riley Pearce! Long time no—'

Riley cut him off. 'Those bastards in the Senate blocked our bill,' she said, with fury in her voice. 'Everything was agreed. Every last bit of detail. Then, at the last moment, something else crops up, something totally unrelated, and our bill is dead in the water. For the last twelve months we've been working with an exposure bill the entire industry had agreed upon. We have to be able to hit the ground running the moment it's signed into law. And now? Words fail me. They absolutely fail me.'

Blake mustered all the empathy he could. 'You must be feeling really frustrated,' he suggested.

Riley sighed. 'Just promise me this, Blake. I have a list of senators who are to be prevented from ever receiving an inGenie. Tell me you'll black ban them for life.'

Blake chuckled. 'Before I can ban anyone, I have to actually prove that inGenie works as specified.'

'I know,' Riley said. 'But I know you well; you work quickly. There's every chance these mouth breathers will still be alive in eighteen months' time.'

'True true.' Blake chuckled. 'But I'm sure their job is stressful. Maybe the stress will take them out. I've just met with Marguerite. We've been having problems with a mystery competitor. She reminded me all projects are political.'

Riley did not respond. Blake counted to five. 'Riley?' he asked.

'I'm still here. What did you just say?'

'*All projects are political.*'

And then the line went dead.

'You seem distracted.'

Blake put down the coin he'd been walking across the back of his hand, and looked up at Taybridge. 'Distracted? That's putting it mildly. It's not that I can't focus. I just can't focus on the right things.'

'Is it Recombinant?'

'Sure. They're still gnawing away at me. But mostly it's Riley.' Blake summarized his conversation with her. 'It's been almost a week, and she hasn't called back. Think maybe I've insulted her?'

'Isn't she your best friend since preschool?' Taybridge asked.

'She is.'

'So I doubt she's just going to cut you off just like that.' Taybridge snapped his fingers. 'Why don't you give her a call? We need you back on board. That coin roll trick isn't going to shape the future of genetics.'

Blake glanced at his phone. He could think of any number of good reasons to delay the call. Taybridge read his hesitancy. 'Can you enter your password and hand me your phone, Blake?' he asked.

Blake surrendered to the inevitable. He unlocked his phone and passed it to Taybridge. It only took Taybridge a second to find Riley under Favorites. He hit the number and passed the phone to Blake. 'Now you'll have to speak with her,' he said.

'Not at all! I wasn't the least bit offended. In fact, I needed you to remind me. When I'm angry, it's easy for me to forget the basics. All projects are political. And we're dealing with politicians. So I just needed to find the right point to exert some leverage.' Riley sipped her coffee and smiled at Blake.

'Sounds like you've already achieved what you wanted,' he said.

'The bill passed in an extended session last night.'

'How did you manage that?'

'Remember the advice Marguerite gave you about playing with Recombinant?'

'Not to sink down to their level?'

'Exactly. Well, she would have given you very different information about dealing with the Senate. She'd have told you to do whatever it takes to push your bill through.'

'Why the difference?' Blake wondered.

'Simple. With Recombinant, you're dealing with an unknown quality. You don't know what cards they're holding, or what hand they're likely to play. This risk of overreach is too great. But with the Senate, we know exactly where we stand. The buck stops with us.'

'With us?'

'With the finance sector. Who controls the funding that ensures those ingrates win re-election? Who pays enormous sums of money to finance their campaigns? We do. As soon as you reminded me of Marguerite's maxim, I knew exactly what I had to do. Sorry about hanging up on you, Blake, but I was on a mission!'

'So how did it all play out?'

'Brilliantly. First, I met with the project managers in all the other major banks. They were as livid as I was. Billions of dollars were hanging in the balance. So we escalated it to the C-level. They green-lighted us to neutralize those hold-out senators. Mission accomplished in less than a week.'

'I'm not sure I want to know the details,' Blake mumbled.

'Everyone has their weaknesses,' Riley explained. 'Collectively, the lobbyists who work in the finance sector know where all the bodies are buried. We targeted the ringleaders first. Once they realized we held all the aces, they folded—and their foot soldiers immediately followed suit.'

'Wow. When we were kids playing in the sandpit together, I never imagined that one day you'd be bending federal politicians to your will.'

'Knee-deep in blood, Blake, knee-deep in blood.' She clenched her fist, and opened it in front of him. 'All the perfumes of Arabia will not sweeten this little hand.'

A few weeks later, Taybridge knocked on the door of Blake's office. 'We have a problem with our drop testing,' he announced. 'Probably nothing major, but the number of errors has increased by 30% over the last month. And when I look ahead on the schedule, we're starting to really push the envelope. If this trend continues, we could blow out our timeframe.'

'Who's the lead on this element of the project?' Blake asked.

'Wilson. He's looking pretty strung out right now,' Taybridge warned.

'OK. Let's go see if we can find the root of the problem.'

Blake and Taybridge crossed the room to Wilson's workstation. Wilson sat at the center of a small bunch of worried-looking coders. 'I think we've found the bug,' Wilson told Blake. 'Although calling it a bug is a bit too glamorous. It's just a syntax error in the code.'

'I don't want to call anyone out—,' Blake began. But before he could finish, a tousle-haired young coder raised his hand.

'It's my mistake,' he confessed. 'I'd been over the code so many times, but somehow it escaped me.'

'OK.' Blake paused for a moment to compose himself. 'We've noticed there are more errors creeping into our work. I guess my question is this: what do we need to do to reduce the error rate?'

The members of Wilson's clique glanced at each other, but none of them dared speak.

'Is it really that bad?' Taybridge asked.

A couple of them nodded. Again, none spoke up.

'What, exactly, is that bad?' Blake asked.

Wilson sighed, and leaned back in his chair. 'Morale is down,' he said. 'Ever since that Recombinant thing. It's been hard for us to focus.'

Blake stared at Wilson. 'But I spoke to the team!' he exclaimed. 'I touched base with Meadows and Marguerite. We reviewed our security protocols and tightened them without going over the top. What more do you guys want from me?'

Wilson winced. 'Blake, you handled that crisis pretty well. We came through it intact. We'd just have liked you to spend as much time with us as you did with your VC buddy and your consultant. You didn't really keep us in the loop.'

'Well, for the love of...' Blake choked back his anger. The pain started to glow on the side of his neck. He turned to Taybridge. 'Did you know about this?' he demanded.

'Why don't we talk in your office?' Taybridge suggested.

'After your last meeting with Marguerite, she called me up,' Taybridge said. 'She summarized your report. Her words? "Sounds too good to be true. What's your take?" I told her everything seemed to be tracking well. She told me to look deeper. "People don't always show their true feelings, she said. Look at the data. They'll give you a real indication of what's happening.""

Blake reached into his top drawer and rifled around. He found a couple of ibuprofen tablets, and swallowed them straight.

'You want a glass of water?' Taybridge asked.

'I'm fine. They're not as hard to swallow as what you're about to tell me.'

'Don't be too harsh on yourself, Blake. I thought everyone was fine. But it turns out they're not. This business with Recombinant has distracted them from the task at hand. And the proof's in the error rate I quoted earlier.'

'That's our problem,' Blake said. 'How do we fix it?'

'Like Wilson said. We spend more time with the team. Find out what's troubling them, and work with them to address their concerns. But we need to do more. We also need to anticipate problems before they arise. So far, we're still reactive.'

'I can't anticipate clowns like Recombinant!' Blake spluttered.

'I know,' Taybridge agreed. 'They blindsided me, too. But anticipating problems is a core part of our job descriptions.'

'So now I look like an idiot,' Blake said. 'Marguerite saw right through me. Wilson and his crew watched me have a total meltdown. What must they think of me?'

Taybridge took a deep breath. 'I know this is painful,' he said. 'But it's not really about you. It's about the team. And they need you to be working at your peak.'

$\mathscr{T}$HIRTEEN

'There are some dark clouds gathering.' Meadows centered his knife and fork on his empty plate and dabbed his napkin against his lips before making this declaration. 'There are forces I can't control. How will it play out? I can't say. But you need to know what's happening.'

Blake pushed his last morsel of fish to one side. 'Are you trying to worry me, Julian?' he asked.

'I want our project to succeed. That means being honest with you if I can foresee any trouble,' Meadows replied.

'You believe there are problems with our project?'

'Not internal problems. I know you've had some hiccups, but from where I sit, everything seems to be on track. Right now, I'm more concerned about external threats.'

'Such as?'

'Private equity companies. Given its earning potential, Med•evolv is underfinanced. Which makes it a likely target. There are a couple of funds sniffing about in the area of high-tech medicine. And at some stage, they'll catch our scent.'

'But our technology is not yet proven!' Blake exclaimed. 'We won't earn a penny if this project doesn't deliver.'

'Not true,' Meadows countered. 'Never underestimate the value of your intellectual property, even if it remains unproven. A corporate raider can monitor Med•evolv's patent applications. They can tell whether you've discovered something worth seizing.

'Same with your staff. Whatever you may think of them, you've assembled a well-credentialed team. They'd be an asset to anyone who believed they could tweak your IP, hype an Initial Public Offering, and turn a profit that was both quick and substantial.'

'I can't say I like the sound of that.' For the first time in weeks, Blake felt the vice beginning to tighten around his skull. 'But you're solid, aren't you, Julian?' he asked.

'Solid?' Meadows stared off into the distance. 'I'm doing everything I can to shore up my position. Whatever happens, I'll be fine, at least from a financial point of view. No one can touch my private wealth, but it wouldn't be enough to save you if it came to the crunch. Not that I'd even offer it. I make it a rule to limit the percentage of my own money that I invest in any one venture. And I've reached that with Med•evolv.'

'So the money you've invested in Med•evolv belongs to other companies?' Blake asked.

'Eighty percent of it,' Meadows confirmed. 'Which is why I need you to hear what I have to say. *Your loyalty is to the project, not to the person.* If you wake up one morning and find things have changed, don't waste your time resenting the new regime. If you feel any loyalty to me, you'll focus on bringing inGenie to the market.'

'You'd want me to buckle under and serve someone who'd deposed you?' Blake asked. 'Surely you'd want me to protect your legacy?'

Meadows shook his head emphatically. 'No. I'd want you to protect my investment. I still have 20% riding on your success.'

Blake nodded. 'My success is your success,' he said. He hesitated for a moment, pressing his fingers against the side of his skull. 'That woman you told me about, your high performance coach...'

Meadows leaned forward. 'Yes?'

'I met with her last week. Like you said, everything I tell her is confidential. But for the first time in my life I'm managing my anger and frustration better. So: thanks. I should have had the courage to see her sooner.'

Meadows smiled. 'My sentiments exactly.'

'Something's not quite right,' Taybridge told Blake. 'I'm sensing some tension between Wilson and Roberto Alvarado, our genetics lead. Not that anything's been said. But they barely acknowledge each other anymore.'

'They need to work closely together at this point in the project,' Blake remarked.

'They do,' Taybridge replied. 'But it's not happening.'

'Let's meet with them. Within the next half-hour. The boardroom's free.'

Taybridge checked his tablet. 'Nothing on either schedule that will excuse them,' he said.

Wilson cleared his throat with a sound not quite brave enough to pass as a cough. 'We may have the smallest of possible problems here,' he conceded.

Blake tilted his head sideways. 'How small?' he demanded.

Wilson glanced at Blake, then Taybridge. He would not look Roberto in the eye. Before he could speak, Roberto cut in.

'In Wilson's case, I believe the correct definition of *small* is *project ending*,' he said.

'Thanks, Roberto,' Blake replied, 'but I'd like Wilson to answer. Before he does, though, please remember that any problems we experience are our problems. We all own them. There's no point wasting energy sheeting home the blame when we could be using that energy to develop solutions.'

'Here's the thing,' Wilson explained. 'We've been challenged by the need to manage extremely complex data sets. Our most effective strategy so far has been to develop a database—codenamed *Double Helix*—utilizing a high degree of data compression. But this fails in one key area: the database is reactive, rather than predictive.'

'Meaning it cannot anticipate likely future changes in a client's genome?' Blake asked.

Wilson nodded. 'If we go with this technology, we'll need to downgrade expectations when inGenie goes to market.'

'Thanks for being so honest, Wilson,' Blake said. 'So what's your view, Roberto?'

'Without a dynamic model, the product falls way short of our vision,' Roberto began. 'I'm disappointed

Wilson's team hasn't been able to translate their thinking into workable code.'

'Take a look in the mirror, Roberto,' Wilson fired back. 'Maybe we'd have more success if you didn't keep shifting the goalposts.'

Roberto bit, and bit hard. 'It's called science, Wilson. Maybe if you were as economical with your coding as you are with the truth...'

'Come on, guys!' Blake stretched out his arms. 'Blaming each other isn't going to change a thing. Just let me know what you need. No finger-pointing. No snarking. Let's focus back on the project.'

'More time,' Wilson volunteered. 'We just need to tweak the code a little.'

'Tell me, Wilson,' Blake asked. 'Is it just a matter of tweaking the code? Or does this require me to take a leap of faith?'

Wilson shrugged his shoulders. 'The solution's not clear cut,' he admitted.

'Come on, Wilson,' Roberto said. 'We have a solution. It's just not the one you favor.'

Wilson glared at Roberto. Blake noticed his anger. 'Tell me more about this solution,' he said to Wilson.

'Why me?' Wilson demanded. 'This is his idea!'

'I'm asking you because you're the one who seems conflicted,' Blake replied.

'I'm more than conflicted. I'm seriously annoyed. Paul Groves, one of my programmers, and Danielle Kai, one of Roberto's geneticists, anticipated this problem way back. Neither Roberto nor I agreed with them at the time. But Groves and Kai kept working on an alternative approach.'

'And?' Blake prompted.

A heavy silence followed. Eventually, Alvarado spoke up.

'Wilson's still angry that Paul and Danielle came up with a brilliant solution,' he explained. 'When I saw what they'd achieved, I felt as jealous as hell. I got over it; Wilson didn't.'

'Is that true, Wilson?' Blake asked.

'Their work is more than brilliant,' Wilson admitted. 'But it could blow out our timeframes and our budget.'

'Weren't you asking me for more time just a couple of minutes ago?'

Wilson pursed his lips. 'I should have thought of it. That's what you pay me for.'

Blake shook his head emphatically. 'Not at all. I pay you to achieve results. Not to block genuine innovation.'

'I hear you,' Wilson said, 'but we're in a gray area from an IP perspective. Groves and Kai worked on this in their own time.' He glanced away, and swallowed hard. 'I let them take the firmware home with them,' he whispered.

'You know that breaches our security policy!' Blake exclaimed.

Wilson turned to Roberto. 'You see why I couldn't speak up?' he said.

Roberto glanced across at Blake. He saw the cold anger in his CEO's eyes. 'Before you start beating up on Wilson, can I suggest you look at Paul and Danielle's work? It's amazing what caffeine and youth can achieve when they work closely together.'

Blake took a deep breath. 'It had better be worth violating our security procedures,' he warned.

Roberto leaned forward. 'These are two young people who want to help you out,' he said. 'Go easy on them.'

'Are they able to speak for themselves?' Blake demanded.

'They're waiting outside,' Roberto replied.

'So bring them in!'

'You're all familiar with the limitations of *Double Helix*,' Paul began. 'Danielle and I have devised an alternative approach which addresses these limitations, while also requiring less operational memory and using an ultra-low power chip. In essence, we've developed a simulation—'

'But we considered a simulation earlier in the process!' Blake cut in. 'The resource implications were unsustainable!'

Paul waited patiently for Blake to finish. 'That's true,' he conceded, 'if you believe you have to build every element of the simulation digitally. However, if you establish a set of mathematical formulas that allow the simulation to build itself...'

'You mean you've developed biological simulation within the computer?' Blake asked.

'Biological and evolutionary,' Paul replied. 'Let us explain.' He pressed a button on his remote, and the first slide appeared on the projection screen.

Fifteen minutes later, Paul and Danielle set down their laser pointers, and waited for questions. An awkward silence filled the boardroom as Blake toyed with his pen.

'You know what?' he said. 'I understand Wilson so much better now. I hate you two guys! That is genius in a bottle! I'm kicking myself I didn't think of that.'

'Well, we couldn't have done it...' Paul began. Then Danielle nudged him, and he stopped speaking.

'Do you have any questions?' Danielle asked.

'Yeah,' Blake said. 'Given how clever this program is, it must have a wicked codename.'

Paul glanced at Danielle. 'We've called it *Fonteyn*,' he said.

'*Fonteyn* as in Dame Margot?' Blake could not keep a note of bewilderment from his voice.

'To us, this software is as elegant as a *pas de deux* between Fonteyn and Nureyev,' Danielle explained. 'Each step leads flawlessly to the next.'

'Makes sense,' Blake said. 'And your logic is impeccable. Two questions. One: where are you in the development cycle? Two: how will it play with our existing software and hardware?'

'We've mapped out the entire program and developed the seed formula,' Paul said. 'All our testing shows it's working as we anticipated. And because we know the firmware you've used in the current prototyping, we can guarantee it will port across without any trouble.'

'Excellent,' Blake said. 'Are there any other questions?'

Wilson and Roberto spoke simultaneously. Blake held up his hands. 'Let's try one at a time,' he suggested. 'Roberto, you're first.'

$\mathscr{F}$OURTEEN

'At first, I felt really angry,' Blake told Riley. 'These two kids—Groves and Kai—had compromised security by sneaking our firmware out of the building, so they could develop their side project on their weekends. But Wilson and Roberto both knew what they were up to. They gave them tacit approval to build *Fonteyn*—but neither of them had the courage to let me in on the secret. If Taybridge hadn't picked up on some rising tension between Wilson and Roberto, I'd still be in the dark. So now...'

'Do you feel angry with Wilson and Roberto?' Riley asked.

'I was. I thought they should have come to me. But then I realized—'

'That despite all the positive things you've achieved as CEO, your senior staff remained in awe of you?'

'What are you, Riley?' Blake protested. 'Some sort of mind-reader?'

Riley shook her head. 'No. Just a project manager. Joining the dots is what I do.'

'So if you were in my shoes, would they have brought you into the loop?'

'Most likely,' Riley agreed. 'But that's not the question you really need answered.'

'OK then. I'll bite. What question should I be asking?'

Riley leaned forward. 'Ask yourself this: *now that I understand what's happened, how will I behave differently in the future?*'

'You mean, *what am I going to change in myself?*' Blake sounded confused.

'Exactly.'

Blake scratched his head. 'I have no idea,' he said.

'Sit back,' Riley suggested. 'Enjoy the rich creamy goodness of your mocha. And let me tell you a little tale.'

Blake chuckled. 'I'll do my best to look interested,' he joked.

Riley gave him her most intense schoolmarm look. 'As well you might,' she said. 'You may remember me telling you how I schmoozed our lawmakers and lobbyists. Of course, this involved a number of transcontinental flights, nights away from home, endless dinners and social functions. You may also remember me telling you that I returned victorious—we won the agreement we'd wanted. However, those weeks of relentless campaigning took a toll on my relationship with my team.

'My first day back in the office, I called everyone together. I'd prepared a little victory speech downplaying my contribution, and exhorting everyone to pull together to finalize the project by the time the legislation was enacted. For my pains, I received a little smattering of applause. I was dumfounded. I'd saved their bacon, and everyone just claps politely? I had expected to be treated as a conquering hero.

'But here's the thing. While I'd been gallivanting about cultivating a new group of key influencers, I'd neglected everyone back in the office. Something in my manner when I returned suggested more than a hint of arrogance. Now as you know, I'm not usually conceited. But power and success went to my head. Even people who I thought were my friends proved a little wary of the new me. It took several weeks of close work to rebuild all those relationships. And even to this day, there are some that haven't returned to normal.'

'Touching story, Riley,' Blake said. 'But what does it have to do with me? I've been with those people every day. I haven't been jetting around playing kingmaker.'

'I realize that,' Riley replied. She tapped her fingers on the table. 'But still, you're the CEO. Everyone on your team knows your status. If you want people to tell you the truth, you need to create a climate in which they feel comfortable approaching you.'

'But I've not done anything to make them feel uncomfortable!' Blake protested.

'Not from your point of view,' Riley conceded. 'But it's not your point of view that matters. It's theirs. And it's not a question of you doing anything to make them feel uncomfortable. It's about you taking positive steps to gain their trust, and ensure they're willing to speak with you.'

'So what should I do?' Blake asked.

'Speak with Wilson and Roberto. Ask them if there is anything you could do differently that would ensure they bring their concerns directly to you.'

Roberto stared out of the window of Blake's office. 'You're right,' he admitted. 'I should have told you of my concerns. Even better, Wilson and I should have presented a united front, rather than sniping at each other. It's weird, you know. We both hoped Paul and Danielle would find a solution, but Wilson was the one carrying most of the risk. He's a coder, so he felt that Paul shaded him. And Wilson was the one who allowed them to smuggle the firmware out of the office. He was reluctant, for sure. But Paul and Danielle were adamant. They needed the firmware before they could write a line of code.

'So why didn't we come to you?' Roberto shook his head. 'We're both hearing some unsettling rumors. That there's a takeover bid looming for Med•evolv. The story we hear is that the Recombinant people are behind the move. So Wilson and I were worried about letting you down.'

'I've not heard those rumors,' Blake replied.

Roberto took a deep breath. 'Maybe there's nothing to them,' he said. 'We've not been able to validate them. But Wilson and I felt a sense of urgency. We wanted to exhaust all the possibilities before we threw our hands in the air and called on Groves and Kai. Plus, we didn't want to distract you. We figured you had your hands full dealing with the external threat.'

Blake laughed. 'I'd have been more worried if I'd heard those rumors,' he said, a note of resignation in his voice. 'But assuming that we weren't facing a hostile takeover, was there anything else that would have stopped you from coming to me?'

Roberto hesitated. 'I'd like to say *no*. But you're the boss. It doesn't matter how friendly the boss acts. He's still different. It's nothing to do with you. It's just that your position is a little intimidating.'

'Fair call,' Blake said. 'But Danielle came to you and told you of her concerns.'

Alvarado shook his head. 'Not really. She let her teammates know, and the word got back to me.'

'So she didn't feel any more confident about approaching you than you felt about approaching me?'

'I'm afraid not.'

Marguerite was working on a project in Europe. It took Blake the best part of twenty-four hours to set up a Skype conversation with her. As soon as the video flickered into life, he began to relax. He trusted her to have the answers he needed.

She waited patiently while Blake explained his dilemma. 'It sounds as though you've simply behaved the way everyone expected you to behave,' she said once he had finished.

Blake exhaled loudly. He felt as though she had let him off the hook.

'But your job is to behave the way people need you to behave,' she said. 'Give them what they need. Not what they expect.'

'You're saying I could do better?'

'I'm saying that you're going to do better. Your project is reaching a critical point. Now, more than ever, your team needs your leadership. And you can't lead if you're inaccessible.'

'Alvarado says I'm a little intimidating. And he says his people feel the same way about him.'

'Exactly,' Marguerite said. 'Everyone expects their boss to be a little distant. But as you've just discovered, that heightens the risks for your project.'

'So how do I change this?' Blake asked.

'Before I answer you, can I check one thing?'

'Sure.'

'Are you comfortable backing Grove and Kai's plan?'

'Well sure, if they tell me it can be done without any extra funding, without pushing our end date back.'

'That doesn't answer my question,' Marguerite insisted.

Blake sighed. 'At first, I felt the same as Wilson. Jealous. Envious. But that won't help me achieve the outcome I desire. So I put my feelings to one side.'

'You've mentioned this to your high performance coach?' Marguerite asked.

'Not yet,' Blake said. 'My next appointment's on Friday.'

'As you probably know, it's important you discuss this with her. Work through any feelings you have that may impede you. Right now, you need to resolve anything that might block you from succeeding.'

Blake reached for a notepad. 'Give me a moment to jot that down,' he said.

Marguerite waited until he had finished. 'Now tell me,' she said. 'Have you informed Meadows about your change of plan?'

'Not yet.'

'May I suggest that you do so? Because whatever Groves and Kai may tell you, any changes could stretch both your timeline and your budget.'

'Meadows is out of town. I can catch up with him early next week.'

'Great,' Marguerite said. 'But remember: tell him what's happened, and explain that it may affect your budget and your milestones. This may throw him a little. Don't argue with him, don't demand, don't manipulate. Instead, give him choices. That will bring him onside.'

'I feel a little daunted,' Blake confessed. 'I don't want him to think I'm failing.'

'That's a useful insight, Blake. Because you understand how Alvarado feels about talking with you, and you understand how his team feels about talking with him. Let's look at how you—and Alvarado—can change that dynamic.'

'I'd welcome any hints,' Blake said.

'First thing is: be up-front. People are nervous confronting you because there's an unspoken law—*don't confront the boss*. These unspoken rules wield a lot of power. So talk with your people one-on-one. Acknowledge you're the boss. But set some clear expectations as well. If you're going to be fully effective as their boss, you need people to come to you when problems arise. Not little everyday problems—they can address them within their teams. But big problems—like this issue with *Double Helix*—these they need to raise with you directly.

'That's the easy part. The second step is harder. When people do come to you, make sure your response is calm, welcoming, and positive. This may not be easy. Your people may be anxious or emotional. They may even be angry with you. If you react badly, you'll shut down the whole experiment. They need you to be open. If you respond well, they'll tell others, and you'll build your reputation with your staff.'

'I thought I handled Damien OK,' Blake said, 'when Recombinant stole his USB stick.'

'Exactly!' Marguerite replied. 'So you've already proven you can do this.'

'Even so, it seems my people are still reluctant to approach me.'

'Building these relationships takes time,' Marguerite said. 'If Damien felt ashamed, maybe he kept that conversation private. But if you'd bawled him out, other team members would have heard about it. Trust me on this.'

Blake nodded. 'Makes sense to me.'

'Finally,' Marguerite continued, 'act on what you've heard. Look into it. Maybe you need to change something; maybe the changes your people suggest won't be effective. Either way, get back to them. If someone has raised an issue with you, let them know what you've decided. If you can't follow through the way they'd like, explain the reasons behind your decision.

'There are few things worse for someone's self-esteem than a blanket refusal, with no explanation given. For some reason, many bosses are afraid to explain why they've chosen a specific course of action. Perhaps they don't trust their own judgment. But you're going to be different, aren't you?'

Blake nodded. Despite the thousands of kilometers between them, he felt the intensity of Marguerite's gaze.

'One last thing,' she added. 'Acknowledge their contributions publicly. When someone comes to you and shares a concern, bring their actions out into the open. Others will follow suit once they know it's safe.'

'And that's all I need to do?' Blake asked.

'Am I sensing some reluctance?' Marguerite demanded.

'It's just that your suggestions seem to run counter to most of the managerial behavior I've witnessed and encountered,' Blake said.

'And how did that behavior work for them?'

'Sometimes, not so well,' Blake admitted.

'If you have any reservations, speak with your coach,' Marguerite suggested. 'She'll help you find the courage you need.'

'I guess I'd expected this,' Meadows said, nursing his macchiato. 'But all projects of any substance experience these dramatic changes of course. I'd say yours is right on schedule. Am I right?'

'There's always a risk the timeline will blow out a little,' Blake said. 'Taybridge is right on it. He's mapped out all the risks and interdependencies. He's riding those charts of his like a rodeo champion. He'll catch any problems before they emerge.'

'You trust Kai and Groves to bring their deliverables in on time?'

'They've achieved a phenomenal amount in the time they've used so far. That gives me faith in the future. The only risk is if one of them falls ill. For the next month or so, they're irreplaceable.'

'I'll burn a candle for them,' Meadows said. Blake could not tell whether or not he was serious. 'You mentioned contingencies. What are your options?'

Blake took a deep breath. 'There are a number of unknowns. Groves and Kai believe they can stay within budget and on time, but Taybridge and I think they're being naive. We've run a ruler over all the figures. At best, we're looking at a 5% budget overrun; at worst, 25%.'

'And the timeframe?'

'Keeping to time will cost more. That's when we creep up to the 25% mark.'

'Because of the extra complexity involved in the programming,' Meadows asked, 'or the need to interface with the sensors?'

'Both. Groves and Kai have demonstrated the kernel of their program, and it works brilliantly. But the more it evolves, the greater the risk—and it's a program that won't meet our objectives if its evolution takes it off track.'

'And there's no other way through this problem?' Meadows asked. 'If the simulator fails, what's your Plan B?'

Blake glanced away. 'There is no Plan B right now,' he admitted.

'Is 25% a solid ceiling on expenditure, or is it a little permeable?'

Blake studied his fingernails intently. 'It's more than a little permeable,' he said.

'You're going to need extra people and extra testing,' Meadows said. 'I'm convinced you're taking the project in the right direction. But the stakes are rising. I'm going to need to crunch some numbers myself. By the way—how much are Groves and Kai asking for the IP?'

'A percentage of gross.'

'Gross revenue or gross profit?'

'Revenue.'

'What kind of percentage are they talking?'

'One-point-five percent each.'

'If I remember well, you're predicting a fourfold return on investment of $200 million over ten years.'

'Correct.'

'Sounds like three points from the gross profit would be more than generous,' Meadows suggested.

'Well, when you do the math...'

'Always do the math, Blake. What have you promised them?'

'Nothing. Said I needed to think on it.'

'Wise. What's their end date? Two years? Five years? When the patent lapses?'

'They didn't say.'

'So this is all an ambit claim. They're making it up as they go along. Would they settle for a share of the IPO?'

'Perhaps.'

'Blake, I appreciate what they've accomplished. *Fonteyn* is going to underpin everything Med•evolv does over the next three to five years. But they can't do anything without us, just as we can't do anything without them. They need our firmware and they need our hardware. I don't mind paying what they're worth, but I have people who will crucify me if I pay them a cent over market value. See if they'll accept payment in stock options.'

'Fair enough,' Blake agreed. 'What kind of limits do you suggest?'

'You value their work, don't you, Blake?'

'Absolutely.'

'You're kicking yourself because you believe you should have envisaged this kind of evolutionary simulation yourself?'

Blake nodded. 'It's so obvious to me now...'

'So you decide on the limits,' Meadows said. 'After all, it's going to come out of your stock allocation.'

FIFTEEN

'Things are going to change around here, sonny, and they're going to change big time!' Lincoln Karcher glared at Blake, daring him to flinch. But Blake refused to play by the rules.

'I'm keen to hear your plans for Med•evolv,' he replied, with genuine warmth in his smile.

The coup had come quickly. Although Meadows and Blake had been expecting it, they were powerless to resist. Money doesn't speak. It bludgeons its way to the front of the queue. Then it bellows.

'You failed at Pyrouette,' Karcher snarled, 'so it's better than even money you'll fail here, too. If you're not part of Med•evolv's future, why would I take you into my confidence?'

Blake leaned forward and lowered his voice. 'There's only one reason why you'll take me into your confidence,' he said.

'As many as that?' Karcher scoffed.

'You'll take me into your confidence because you want a 400% return on your capital over the next ten years.'

Karcher burst out laughing. 'Why so little? If you're going to promise the world, why not throw a few stars into the bargain?'

'There are some coat hangers over in the hutch behind reception,' Blake said.

'What are you on about, sonny?' Karcher demanded.

'Hang up your jacket. I'll take you through the lab.' Blake waited while Karcher fumbled with his coat. 'Once you've seen what we've achieved,' he told himself, 'you'll be laughing on the other side of your face.'

The evening before, Blake had met Meadows for a quiet drink. The downtown bar was nearly deserted. Meadows ushered him into a discreet corner booth.

'I know you're disappointed,' Meadows said.

'It's just business,' Blake said, his jaw set hard. 'You had a chance to cash out, and you took it.'

'My partners and I came to the only possible conclusion. When we balanced the risks against the returns, the percentage of future earnings Karcher offered was impossible to refuse.'

'He's injected some much needed extra capital,' Blake admitted. But I've conducted my own investigation. He's a racketeer, an asset stripper, a corporate raider. I can't believe he wishes us well.'

'Perhaps not. But you're best to approach this new relationship from a position of strength. Tell me—who's your favorite actor?'

'Christian Bale.'

'Name a film of his you liked.'

'*American Hustle*.'

'What did like about it?'

'Fast-paced comedy that keeps you guessing. There's a con within a con within a con. And Bale and his comb-over are just brilliant.'

'Co-stars?'

'Amy Adams. Bradley Cooper. Jennifer Lawrence. Oh, and de Niro.'

'Who financed it?'

'What do you mean?'

'Movies like *American Hustle* cost plenty. Maybe forty or fifty million. Who ponied up the cash?'

'I have absolutely no idea,' Blake admitted.

'Nor do I,' Meadows said. 'But without them, Irving Rosenfeld and his comb-over would never have entered your consciousness.'

'So you've seen it, too!'

'The more I learn about con-artists, the more I learn about business,' Meadows replied dryly. 'But you're missing the point. Everyone recognizes the actors, the drama and humor of the story. No one remembers the money man.'

'And Karcher is the money man,' Blake muttered to himself.

'Every money man is looking for a sure bet,' Meadows said. 'You're that sure bet. Karcher wouldn't have offered my partners and me such a generous payout if he thought you'd fail. Mind you, he'll use every trick in the book to destabilize you, to chip away at your self-esteem. Stay strong. In ten years' time, when people can anticipate and forestall cancer, they'll thank Med•evolv. They'll thank Blake Stein, not Lincoln Karcher.'

'I'll be straight with you, Blake,' Karcher said. 'I don't want no sanitized tour. I just want ways to cut costs and maximize profits.'

'I'd be disappointed if you were asking for anything less,' Blake replied. 'I'm taking you right into the heart of the skunkworks. You'll meet the team that's building inGenie from the ground up.' Then Blake propped, and turned to Karcher. 'This isn't really what you want, is it?' he asked.

'What do you mean?'

'These people are just overheads! We really need to trim some of the fat.'

'Are you having me on?' Karcher growled.

Blake looked confused. 'You said you want to save money...'

'Just show me what you got,' Lincoln demanded.

'No problem.' Blake gave Lincoln a big smile. 'These people are the best in their fields. But don't worry. As I have long since discovered, there's no such thing as a stupid question.' He pushed the door open, and ushered Karcher into the workshop.

'Holy snapping turtles!' Karcher exclaimed, pointing at the whiteboards that covered the walls. 'What is this? Modern art?'

'Lincoln, I'd like you to meet Aiden Taybridge, our project manager. He can interpret the hieroglyphs for you.'

Taybridge extended his hand. 'A pleasure to meet you, Mr. Karcher. You've noticed our dynamic project plan? Let me show you how it works.'

'Don't bother,' Karcher replied. 'I've never seen a project like yours come in on time and under budget. How far has yours blown out?'

'An excellent question,' Taybridge said. 'Six months ago we switched our core IT technology from a database to a simulator. Our costs blew out badly for a month, and we planned for a slight overrun on our timeline, but we've since refined the entire project. We're back on time, and we've limited the budget overrun to 10%.'

'So what have you cut from your original design?'

'Nothing. We've added five new functions that will enhance its efficacy.'

'So you were overfunded from the beginning, by the sounds of it.'

Taybridge chuckled. 'You know our revised funding envelope. $220 million. And you know what we've promised. So you be the judge.'

Karcher sighed. 'OK. Show me what you got, and how you got it. But I gotta warn you, Taybridge. I don't like your attitude.'

'Then you may not like my modern art, either,' Taybridge said. 'So let's start with the inGenie prototype. Groves and Kai can walk you through its functions.'

Half an hour later, Karcher wiped the sweat from his forehead. Paul and Danielle had answered all his questions in simple language, taught him to use the interface, and had begun building his genetic profile. As much as he refused to admit the truth, his mind was reeling. 'You might have something here,' he grumbled. 'But is gene tech really what this country needs? The human biome is the new frontier. This won't catch every disease.'

'An excellent observation,' Groves said. 'The beauty of the *Fonteyn* engine is that we can use the same core technologies to build simulations of the human biome. Indeed, working with the genome has taught us everything we need about building a more responsive and dynamic simulation that will give us an integrated health reading for all users in real time: genome and biome, and whatever bacterial and viral vectors we can program in as well.'

'What's the status of the biome simulator?' Karcher asked.

'We have three interns mapping out the core capabilities right now,' Taybridge said. 'Kai and Groves are eliminating all the bugs from inGenie Mark 1, to be ready for our launch in two months' time.'

Danielle immediately corrected him. 'Seven weeks, five days, and nine hours, to be exact!'

Karcher mustered the last of his old cynicism. 'That takes two months?' he griped.

Taybridge smiled. 'At this point, we've ironed out 95% of the bugs. That last 5% are the hardest to eradicate. Mind you, with this project, the number of bugs we have encountered is down around 70% on the norm.'

'Seventy percent on the average per ten thousand lines of code?' Karcher could not hide his astonishment. 'Now you got my attention. Show me how you did that.'

'Of course.' Taybridge's smile expanded further. 'Please step into my art gallery.'

'The answer is simple: it's magical,' Taybridge began. 'Real magic. Not smoke and mirrors. And the source of that magic? Even simpler. It's in our people. But it's not what you think.'

'You don't know what I think,' Karcher protested.

'Let me guess,' Taybridge offered. 'You think we've only employed geniuses? The cream of the cream?'

'Something like that,' Karcher grumbled.

'We employed the best,' Taybridge admitted. 'But a roomful of Michelangelos would be a nightmare. Individual genius is nothing. Team genius is what matters here.'

Again, Karcher tried his hardest to disguise his interest. 'Team genius? You mean a day of trust-building exercises, and then you pat yourselves on the back?'

'Three days, actually. But no high rope work, no falling backward into each other's arms. Blake focused on cultivating real trust with each member of the team. We built a collective vision for our success, then Blake gave us the resources we needed to develop inGenie. What followed was nothing short of miraculous. Every team member—from the most junior to the most senior—took full responsibility for his or her own performance. We've all been juggling multiple tasks, myself included. If an intern programmer needs my full support to achieve an outcome the rest of the team is relying on, then I give him 100%. It's my role to serve him.'

'You got that ass backward,' Karcher said. 'Junior staff work for their bosses. Not the other way around.'

Taybridge rubbed his chin. 'Sounds like we have a difference of opinion,' he said, 'and I know I'm not going to convince you. But if you had a junior staff member who asked you to pay for a three-day training course that would take his coding skills to the next level—and pay for itself many times over with a more efficient piece of software—how would you respond?'

'My people know better than to come to me rattling their begging bowls,' Karcher said. 'I'd sack him and find someone with the skills I needed.'

Taybridge smiled. 'This is not about money,' he thought. 'It's about status. Karcher's the kind of executive who would sooner kill a project than offer an ounce of generosity, or admit any weakness.' Wisely, he kept these thoughts to himself.

'I understand your logic,' he said instead. 'We've created a slightly different culture here. Earlier, you said you were impressed with our achievements.'

'Sure,' Karcher replied, 'but not if it means bowing down to the office boy.'

'I'll begin again,' Taybridge offered. 'Without mentioning the interns. Deal?'

'Just make it snappy,' Karcher demanded.

'Of course. I'll summarize what we learned from the consultant who guided us. *One: 85% of projects fail.* That's a lot of money down the tube.'

'As long as it's not my dough, I don't care.'

'But sometimes it is your money. Am I right?' Taybridge read Karcher's face perfectly. He's lost millions on projects that had failed, but would not fess up.

'*Two: there is only one reason for a project. To create something both essential and exceptional.* No more penny-ante projects. No more busywork. Just get on with changing the world.'

'And making me plenty in the process,' Karcher added.

'Absolutely,' Taybridge said. 'High risk equals high return. But Marguerite's next point really annoyed me at first. It felt like a slap in the face. *Three: processes neither make nor break a project. People do.*'

For the first time, Karcher gave Taybridge a look of recognition. 'You're the same as me. You expect people to follow the bouncing ball, to do as they're told. And when they don't follow the rules, you arc up!'

Taybridge nodded. 'In the past, that's exactly how I've behaved. And if I'd done the same here, this project would have collapsed months ago.' He leaned forward and looked Karcher directly in the eye. 'This project was riddled with uncertainty. We wouldn't have pulled through if people hadn't been focused on achieving a common vision, if they hadn't shared their experience and knowledge freely. Which leads to my next point.

'*Four: a vivid, shared vision unites a team.* We brought everyone together. Heard Blake explain what he imagined inGenie would achieve. A truly inspirational speech. Everyone bought into that vision.'

'Even your interns?' Karcher sneered.

'I promised not to speak of them,' Taybridge reminded him. 'But we needed to have everyone on board.

Everyone. So all our team members have their eyes on the end goal. Which means they'll go that extra mile to help each other out. Because if they don't, Med•evolv will fail.'

Karcher tapped his index finger against his upper lip. 'I don't know, Taybridge. It all sounds very Pollyanna. Too much sweetness and light. Doesn't it ever tire you out?'

Taybridge gave Karcher his most piercing stare. 'This has been the most challenging project I've worked on in my entire life,' he said. 'It's so much easier to manage a project by simply following a stipulated project management methodology. So much more reassuring to pretend that everything's under control, that no miracles are required to bring the project home. So much easier to deal with communication problems by sacking staff, rather than listening to them and mediating their disagreements. But you know what? After experiencing this approach, I'm hooked. I'd never try to run a project the old way again.

'Let me give you a practical example.' Taybridge took Karcher over to one segment of the whiteboard. 'Leonard is one of our programmers. Here's a sub-project he needed to complete by the 23rd of March. *Finalize code for Segment G of* Fonteyn. He was writing an algorithm to differentiate between a number of possible mutations, based on a specific chemical signature. Two weeks before the 23rd, Leo came to me. Said he was struggling with the algorithms. He felt there was a risk he might fail.

'Now this didn't surprise me. Make no mistake, Leo is a brilliant programmer. One of our best. But what we were asking of him really pushed the envelope. So once he's approached me, what do I do?'

'Sack the little nerd?' Karcher suggested.

'Incorrect! I asked him who he felt could best help him. He nominated Rachel. We explored the idea briefly, I checked with Rachel's team leader, and the next day Rachel and Leo spent the morning nutting out the problem. That's all it took. It's not that Rachel's smarter than Leo; she just sees problems from a different point of view.

'Now on the face it, there's nothing particularly impressive about this. A few hours to help Leo conceptualize the deep structure of his algorithms. But look downstream at the alternative.' Taybridge walked Karcher along the wall. 'Here, here and here are some critical gates. If Leo fails to deliver on the 23rd, we miss this gate, then this one, and then this. The project pretty much collapses at this point. Multiply that problem by hundreds of other critical interdependencies, and you can see why I'm making such a big deal about this approach.'

'So snot-nosed kid puts his hand up and asks mummy to help,' Karcher objected. 'I still don't see why you're making such a fuss.'

'Because I've worked on many projects where someone like Leo has kept quiet, because they're too frightened to admit their inadequacies. Leo is shy, and he's a perfectionist. Speaking up was hard for him. He only managed it for one reason.'

'And that was?'

'He'd seen others come forward, and be rewarded, rather than punished. That gave him the confidence he needed to assert himself.' Taybridge paused, and smiled thoughtfully. 'In the past, I'd probably have criticized him, made him feel small. Then I would have blamed him if the project fell in a heap. What would you have done?'

Karcher made a fist of his left hand and pressed it against Taybridge's chin. 'Like I said earlier, Taybridge. I don't like you. And now I have extra grounds for my dislike. Don't make things worse for yourself.'

Taybridge reached up, clasped Karcher's wrist, and moved his hand away from his face. 'There's another thing I'd forgotten to mention. It's the opposite of sweetness and light. *Five: all projects are political.* If you're going to step in and lead a project, you'd better be a consummate politician.'

'Not just a headkicker?' Karcher asked.

Taybridge could not tell if Karcher was being serious, or cracking a joke. 'If you kick heads, how long before your people stop trusting you? And if you think they'll work harder because they're terrified, you're wrong. These people are leaders in their professions. Beat up on them, and they'll go and work for a competitor.'

'Sounds to me like you're enjoying your end of the conversation a little too much,' Karcher warned.

'If I can't be honest with you, there's no point in me leading this project. But believe me, for the first time in my life I understand the politics of project management. I'm here by the grace of Blake, and now by your grace as well. So you might think I'll do everything in my power to please you. You'd be wrong. This project won't succeed unless I speak the truth to you—especially if it's something you don't want to hear. The moment my courage fails me, we're done.'

Karcher glared at Taybridge. 'Just remember,' he growled, 'you're far from indispensable. In fact, I've been asking myself why this project needs both you and Blake. Seems we could save some decent cash if we dispensed with your services.'

'You could indeed save some money,' Taybridge agreed. 'Would you achieve the same outcome? Hard to say—if you sack me, how will you ever know? But let me tell you this. If you fired me now, I could walk into one of half a dozen project management jobs tomorrow. Before Med•evolv employed me, I'd break into a cold sweat at the thought of losing my job. That's a change I can take to the bank.'

Karcher stared closely at Taybridge. 'You got a lot of spirit, Taybridge,' he deadpanned. 'More spirit than you could ever possibly need. Maybe I'll keep you—for the moment.'

$\mathscr{S}$IXTEEN

'Two years ago I told some friends of mine about a young woman called Gina. Yes, it's a love story—but not the kind of love you might be thinking of. Gina has been waiting for me for the last two years, and during that time her expectations have grown. Let's see how she's been doing.'

Blake stepped to one side, and a video presentation began on the big screen behind him. A young woman dressed in yoga gear spoke directly to camera.

'Hi. I'm Gina,' she began. 'Blake has probably told you about me. There's one thing that matters to me more than anything else in my life. My health. And Blake promised he'd help me manage my health in ways that have been impossible. Up until now.'

The camera pulled back, revealing the inGenie unit sitting to Gina's left. 'This machine is pure magic,' she continued. 'It contains my entire genome. Not some lifeless database, but a digital 3-D model of my human essence.' She pressed a button on the side of inGenie, and then held up her tablet. A three-dimensional image of a double helix appeared on the screen.

'The women in my family have a history of breast cancer,' Gina said. 'My grandmother died from breast cancer, and my mother had both her breasts removed before she turned forty. A double mastectomy. It robbed her of her womanhood—and failed to save her life. She died of metastasized cancer at the age of forty-four.

'This gene right here renders me vulnerable, too.' She pointed at the screen, where a particular gene was highlighted in red. 'So I want to know it's behaving itself. Up until now, that was impossible. As you might imagine, this has given me many sleepless nights. But inGenie has changed the game.' Gina paused, and held up a circular piece of cardboard, the size and shape of a communion wafer. Discreetly, she ran it down her tongue, and then inserted it into a slot in the front of the inGenie. A green light on the front of the machine began to blink.

'InGenie knows my DNA better than I do,' Gina continued. 'As soon as I insert the wafer, inGenie looks for any changes that signify an increased disease risk. Which lets me sleep soundly—for the first time in my life.'

The screen faded to black.

'I let Gina speak because she has a story worth telling. Let me be clear: Gina is not an actor. She's a real life young woman. Her family history of breast cancer is real. Her fears are real. And now they've been laid to rest.'

Blake stared out at the crowd that filled the 1500 seats in the Capitol Theater. 'Let's recap some of inGenie's key features,' he said.

Twenty minutes later Blake set his remote pointer down on the lectern. 'There's a joke in project management that runs something like this,' he said. 'When you look at the spaghetti tangle of boxes and triangles that litters a typical project management chart, the penultimate box reads: *And here, a miracle occurs.* I had a great team working with me, so we didn't have to rely on that miracle. There's a simple reason why. Miracles, in my experience, are rare. We didn't want to spend that miracle ourselves. We wanted to reserve it for you. And when you plug in your inGenie for the first time, you'll find yourself in the presence of a miracle. Let me promise you this. You're going to love inGenie even more than Gina does.'

As one, the audience stood and applauded. The noise of their hands clapping drowned out the sound of Blake's racing heart.

Backstage, friends, colleagues and wellwishers joined Blake in the crush around the canapé table. Karcher was the first to greet him.

'Great launch!' he said as he slapped Blake on the back. 'I never doubted you, not for a moment!'

'Thanks, Lincoln. You going to stay for my announcement to the team?'

'I gotta rush. But I like what you're doing. We're going to dominate this industry, I tell you. Dominate!' Karcher shook Blake's hand, and pushed his way toward the exit.

'I never regarded you as a born diplomat,' Riley said as she moved effortlessly in beside Blake. 'Which makes your achievement with Karcher even more meritorious.'

'I'd never have learned how to manage someone like him without your counsel or Marguerite's, or without the support of Taybridge,' Blake admitted. 'But once I realized that all his bluster stemmed from a deep sense of insecurity, I found it easy to relate to him.'

'You all set to announce your next project?'

'In a moment.' Blake glanced across the room, where he caught a glimpse of Meadows. 'There are a couple of people I need to thank first.'

Meadows passed through the crowd, and clasped Blake's hand warmly. 'Unlike my evil nemesis Karcher,' he admitted, 'I doubted you. Not that I doubted your vision, just your ability to bring it to fruition. I don't quite know how you managed it, but you've achieved something world-changing. But that's not all. You've changed too. All that pride you used to exude—it's fallen away. And the man underneath is self-assured and genuine. Thanks for giving me the opportunity to invest in you.'

Blake felt a tear forming in the corner of his eye. 'Thank you for having faith in me,' he replied. 'But if you want to thank anyone, thank Riley. And Marguerite. And even my therapist. Without them, I'd never have brought inGenie to market. You know, I thought a project was just a project. I had no idea of its real purpose—to turn me from a self-absorbed savant into a real leader.'

'It's their feminine influence that made the difference,' Meadows said.

'How do you mean?' Blake asked.

'All the rigidity we've imposed on project management over the years has been disastrous,' Meadows replied. 'I've seen it in other start-ups I've backed. Those project planning methodologies shut down communication, they block connection, they inhibit intuition. Riley and Marguerite have helped you balance everything out. They know you can't just throw away the structure. That would be absurd. You have to replace it with something more organic.

'You and Taybridge and your team leaders have done exactly that. You've achieved the right balance between structure and flexibility. Between the masculine and feminine. Between focusing on the outcome, and focusing on the journey.' Meadows gave Blake a rueful smile. 'If only my other start-ups were as self-reflective and as open as Med•evolv. Unlike Karcher, I'm long past wanting to dominate the world. And I believe you are, too.'

'You're right,' Blake said. 'And relinquishing the fantasy of global domination has a powerful side effect. I'm experiencing fewer migraines, and they're much less severe. But you know what delights me even more?'

'You feared losing your creativity,' Meadows replied. 'You believed your migraines heralded your breakthrough ideas. But from where I'm sitting, I see more breakthrough thinking, not less.'

'Exactly. I'd convinced myself it was true, to justify the pain. But now I realize I needed those migraines like a hole in the head.'

'Maintain your current trajectory,' Meadows said, 'and you'll achieve more than you could ever dream.'

'Thank you, Julian,' Blake said. 'And you're on board for Stage Two?'

'Absolutely. I'm looking forward to your announcement.'

'Let me quickly touch base with Taybridge.' Blake shook Meadows' hand again before turning away.

'You're excited?' Blake asked Taybridge. 'Because I'm pumped. I thought inGenie was magic, but our next steps...'

Taybridge could not help but smile at Blake's enthusiasm. 'This time, you're bringing the team along right from the start.'

Blake shook his head. 'It's going to be your team, Aiden. And you're the one who'll be inspiring and guiding them.'

Taybridge looked Blake squarely in the eye. 'I'm as ready as you are,' he said.

Blake stepped up onto the temporary podium erected backstage, and cleared his throat. 'If you can all give me your attention for a moment,' he began.

Everyone set their plates aside, and gathered around. They had heard Blake would be making a second, private announcement after the launch, and they all wanted to hear what he had to say.

'Today is not the end of one project, so much as the beginning of three more,' he declared. 'We all understand the critical role the human genome plays in regulating our health. By focusing so strongly on the genome, we run the risk of misleading our clients into thinking we have everything covered. But we don't.

'The human biome—those millions of bacteria that play vital roles within our bodies, and whose interactions affect our health in ways we don't yet fully understand— is the next frontier. The *Fonteyn* engine that Paul and Danielle developed has the potential to simulate the human biome as well as the human genome. It also offers us the capability needed to explore the interaction between the human genome and the human biome.

'This is an exciting new frontier. And inGenie has rendered the legislation governing this frontier obsolete. So today, I'm announcing not one but three new projects that Med•evolv will be progressing, starting tomorrow. And because you've all made an exceptional contribution to the success of inGenie, I'm guaranteeing every one of you a job on the project that best matches your skills.

'First, we begin work on inGenie Mark 2, codenamed *Nureyev*. This will track and predict the health of the client's inner biome, as well as their genome. We're expecting to see the first iteration within twelve months.'

Blake paused as the audience gasped audibly. 'Yes, I know that timeframe is tight. We also know that other researchers are probing the same field. We have a head start, and we need to maintain our advantage. We've proven we can bring a brilliant product to market in two years. And we've proven we can manage these high risk projects far more effectively than anyone gave us credit for at the start.

'Aiden Taybridge will be leading *Nureyev*. And Wilson will be his understudy, soaking up all Aiden's knowledge, and building his own capabilities as a project manager.

'Second, we're partnering with Bank Pacific West to prepare a model set of legislation that addresses and anticipates all the issues involved with governing the digital biome. As you know, Bank Pacific West has become a significant investor in Med•evolv. The bank knows how to influence the passage of legislation through government, both here and overseas. Riley Pearce will lead that project, codenamed *Choreograph*. This project gives us a unique opportunity to define the field of digital health, not just by creating cutting-edge projects, but by contributing to the development of cutting-edge public policy, as well.

'Third, I'm announcing the establishment of a research center to explore possibilities in digital health, far ahead of any commercial potential. This research center is being set up under a grant we've received from the Meadows Family Trust. Roberto Alvarado will lead a small multi-disciplinary team to really challenge conventional thinking and push the boundaries of health research.'

Blake lowered his voice. 'I know what you're thinking. In the past, there have been pure science research centers that have failed to commercialize the opportunities that fell their way. Med•evolv won't fall into that trap. As CEO, I'll oversee everything. And Riley, Aiden, Roberto, Wilson and I will be meeting regularly to make sure there's nothing we've missed.

'Before I finish, I need to make a confession. A couple of years ago I had a dream, and a sense of unjustified optimism. I failed at Pyrouette. There was every chance I could fail equally at Med•evolv. But I found people who believed in me, believed in my dream, and together we pushed through. I could not have done it without you.

'Today, I have multiple dreams. I've dropped that old arrogance that rubbed so many of you up the wrong way. I don't know quite how we'll achieve these dreams, because the path is uncertain. But I do know one thing. We've learned—all of us, not just Aiden and Wilson and Roberto and me, but everyone in this room—how to manage a project. Now, we're going to take all that we've learned, and apply it to running three projects—projects that, it just so happens, are going to change the world!'

After Blake had been showered with applause, he stepped down from the podium. Then Wilson sidled up to him. 'Our IT security team picked up a security vulnerability that only arises once inGenie is connected online,' he whispered. 'It's not actually a vulnerability in inGenie, so much as a problem with routers running a certain version of firmware.'

Blake ushered him aside, into a private alcove. 'Where are we with a security patch?' he asked.

'It'll go live in twelve hours, with an OS upgrade by the end of the week.'

'A chance to test our notifications to users?'

'Exactly. We're sending email and text messages to all our beta testers warning of the vulnerability. They can choose between an automatic upgrade, or installing the upgrade manually.'

'There may be other routers with similar vulnerabilities,' Blake said.

Wilson nodded. 'That's the reason for the OS upgrade. To tighten security while we test for those weaknesses.'

'If there's any hint our clients' personal health information could be compromised or leaked, then we'll lose their trust on a massive scale. And I don't need to tell you how that could affect Med•evolv,' Blake warned.

'You sure don't!' Wilson replied. 'I'm moving the security team to a high priority operational status, starting tomorrow. Now that we've launched inGenie, security is business as usual. Mind you, we'll need to stay ahead of the game.'

'Absolutely,' Blake said. 'So now we can move the project from pre-launch to pre-operational?'

'And once inGenie hits the shelves,' Wilson replied, 'we move to full operational status. Which will mean project completed!'

For further information and to contact the author:

Email: ddromgold@rncglobal.com

Web: www.rncglobal.com

Diane Dromgold

Diane Dromgold

www.ingramcontent.com/pod-product-compliance
Lightning Source LLC
Chambersburg PA
CBHW031325210326
41519CB00048B/3135